Why Does My Cat Do That?

Why Does My Cat Do That?

Comprehensive answers to the
50+ questions that every cat owner asks

Ivy Press

First published in 2008 by
Ivy Press
210 High Street, Lewes,
East Sussex BN7 2NS, U.K.
www.ivy-group.co.uk

British Library Cataloguing-in-Publication Data
A catalogue record for this book is available
from the British Library

ISBN: 978-1-905695-76-8

The questions and answers in this
book relate to individual cases and
should not be used as a substitute for
a veterinarian's advice. Please see your
veterinarian if you have any concerns
about your cat or its behavior.

Ivy Press
This book was conceived, designed, and
produced by iBall, an imprint of Ivy Press.

Creative Director Peter Bridgewater
Publisher Jason Hook
Editorial Director Caroline Earle
Art Director Clare Harris
Senior Editor Lorraine Turner
Design J.C. Lanaway
Concept Clare Barber
Illustrator Michael Chester
Consultant Veterinarian Dr. Shawn Messonnier

Contents

Introduction

The cat has been a domestic companion to humankind for at least four thousand years. Down the centuries, cats have delighted us with their independent spirit, their animal grace, and their sleek, self-assured beauty—so much so that the cat is now the most popular household animal: In the United States, for example, there are 69 million pet cats, compared to 62 million dogs.

But for all our long familiarity with this lovely creature, the cat remains something of a mystery to most of us. Felines have an otherworldliness, a way of being that seems somehow beyond our human ken. Every cat owner has seen instances of this typically feline behavior in action, and has wondered from time to time, "Why does my cat do that?" What, for example, is the psychological or biological explanation for the way a cat winds itself around our legs in greeting? Why does it have the habit of turning its back on its owner after a scolding? And why does an otherwise friendly, domesticated animal seem to take a rather unappealing delight in tormenting its prey? Cats are unpredictable friends, and their behavior can both fascinate and perplex us.

This is the nub of our long-standing relationship with cats. Our fondness for them—and proximity to them—lead us to believe that their perception of the world is not so far removed from our own. In other words, we see them as furry four-legged humans rather than as descendants of wild animals. This is a misunderstanding of the cat's essential character, and it can cause the most well-meaning and loving of owners to treat their cats in ways that may engender distress or anxiety in the pet, and disappointment in the owner.

This book answers more than fifty questions that go to the heart of cat behavior. The answers draw on the latest thinking of feline psychologists. They will help you to understand your cat, and give you a chance to see things from the feline point of view. If your cat could read, this is the book he would choose for you.

Chapter one
From kitten to adult

Kittens are undeniably appealing, but like all youngsters (both animal and human), they need particular care and attention. **It is vital to get your relationship on the right footing from the very start,** so that your kitten learns to adapt to your home—and you learn how to adapt to your kitten. Many new owners find themselves baffled, for instance, by their kitten's tendency to hide away, to indulge in play-fighting that is a little more fight than play, and, of course, to spurn the litter box when the seller had insisted that it was perfectly housebroken. This section deals with the teething troubles that owners may face when they welcome a young cat into their family. Most of these problems are easily solved with a little love and patience. Before long, your kitten will become a companionable and enjoyable presence in the home.

Why does my cat…
move her kittens?

Q My cat is pregnant with her second litter. Last time she gave birth under the bed in my guest room, but she moved the kittens to the laundry room when they were a few weeks old. I think perhaps we disturbed her by going in too often when they were little—though we did our best to give her as much privacy as possible. What is the best way of making her feel secure this time?

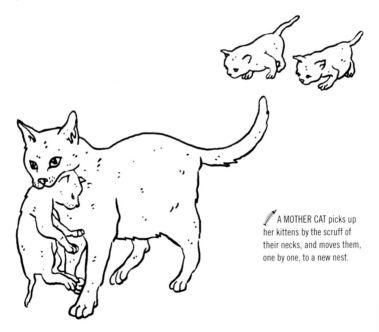

A MOTHER CAT picks up her kittens by the scruff of their necks, and moves them, one by one, to a new nest.

A It is true that a mother cat may move her kittens to a new home if her nest is frequently disturbed. This is her natural response to what she perceives as a threat to her kittens. But this is unlikely to have been the reason why your cat moved her litter. You were careful not to intrude too much and, in any case, she knows you well. Moreover, she had a clean, dry spot in which to look after her kittens: just what she needed.

So it is more likely that your cat was acting on her natural instincts rather than responding to any disturbance. Regardless of any threat, almost all mother cats move their kittens to a new home when they are between three and four weeks old. You see this in the wild, and in domestic situations, too.

It's often suggested that cats move their kittens because they have either grown too big for the birthing area, or because the nest has become dirty. A more convincing theory is that the move has to do with the kittens' developing needs. When kittens are four to six weeks old, they start eating solid food. This means that the mother cat needs a nest that is near to her hunting ground. So, she moves the kittens in preparation for this new stage of the weaning process. Your cat clearly doesn't need to hunt for food, because you provide her with it. But in this instance, as so often with domesticated cats, her wild instincts come to the fore and she reverts to the behavior of her ancestors.

KITTENS ARE REARED entirely by the mother, who grooms, feeds, and protects them. Later on, she teaches them the skills they need for independent survival.

SPECIES WATCH

While nursing cats move their nests, other animals stay put. The female hornbill makes her nest in the cavity of a tree or rock, then literally walls up the entrance with mud so that she is trapped on the inside. She remains immured for months until her chicks are hatched. During this time, she is fed by the male through an aperture in the wall.

Why does my kitten...
pounce on my ankles and bite me?

Q I brought my kitten home when he was seven weeks old. He was quiet at first, but seemed to settle in. But now he has become really boisterous. Whenever I walk past, he shoots out of his hiding place and pounces on my ankles. I didn't mind at first, and was happy to indulge his rough play—but then his scratches started to hurt. He has started biting me, and has drawn blood on several occasions. He is now 12 weeks old, and shows no sign of calming down. How can I make him realize that playtime shouldn't hurt?

A TINY KITTEN'S play-fighting is fun, but a growing cat's ambushes become less appealing once its claws and teeth sharpen.

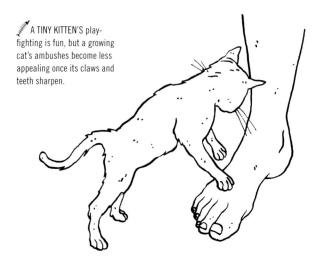

A This is quite a common problem, and it stems from a mismatch between your kitten's feline needs and your ability as a human to meet them entirely. If your kitten had been left with his mother and siblings for longer, he would have spent much of his time play-fighting and play-hunting. His cat family would have let him know—through sharp, reproving nips, or simply by interrupting the playtime—when he was being too rough, and he would quickly have learned his lesson.

You haven't been bothered by the claws and teeth of a tiny kitten, so you've let the behavior pass. To him, this means that scratching and nipping is allowed. But as he has grown bigger, his teeth and claws have become a lot sharper, and capable of inflicting real damage. Also, as you have seen, your little kitten has come to associate you with fighting—which means that the sight of you can trigger an attack.

Telling your kitten off won't help. The best way to stop him ambushing you is to stop making this enjoyable, and to introduce him to lots of other ways to play so that he doesn't get bored. So, ignore his attacks as much as you can. Meanwhile, provide him with plenty of stimulation. Try a "fishing rod" toy—one that is attached to a string and long stick—which allows you to play with him while keeping your hands and body well out of the way. You may also want to consider getting another cat to keep him company (*see page* 124).

SPECIES WATCH

Lion cubs, unlike kittens, learn to hunt by observation rather than practice. They are much smaller than their prey, so cannot at first take part in a real hunt. So they watch from a distance, acquiring a hunter's know-how "theoretically." By the time a cub is two, it is big enough to apply the knowledge, and can bring down a gazelle on its own.

TEACH YOUR KITTEN to play games that don't involve attacking parts of your body.

Why does my kitten...
paw the floor by its bowl after eating?

Q I have a three-month-old kitten who has a funny little habit of pawing the floor beside his food bowl. He does this after every mealtime, but I haven't seen him doing it at other times. He looks as if he is trying to dig for something. I am keeping him indoors, and he is an only cat, so this is not something that he has picked up from a fellow feline. Can you cast any light on this strange quirk?

IN THE WILD, a cat will bury its leftovers so that the smell does not attract the attention of predators. For the same reason, it also buries its feces.

A One of the most fascinating things about living with cats is that they occasionally do things that give you a glimpse of their ancestry. In the wild, thousands of years ago, cats would bring their prey back to the nest and devour it there. In order to stop the scent attracting other animals, cats would bury any food remains. **Your cat will never have to hide his food from predators, but the ancient instincts that motivated his forebears are present in him, too,** and some primal urge drives him to replicate this obsolete behavior pattern. In some cats the urge to "bury" leftovers is so strong that they will drag some suitable item—a towel, say, or a newspaper—to the food bowl and drape it over the food. Taking the bowl away after the cat's mealtimes is often the only way to stop this kind of behavior.

The food area is naturally a central point in a domesticated cat's life, and can be the subject of other seemingly unfathomable behaviors, too. For example, quite a few owners find that their cat regularly leaves a small toy or some other object in the food or water bowl. This is also a throwback to life in the wild. Deprived of the opportunity to hunt for real-life prey, an indoor cat will stalk a toy or ball of paper provided by its thoughtful owner. As the cat does not have a nest to which to take this surrogate prey, it will bring it back to the place where he usually eats: It is the nearest thing he has to a cave.

THE FOOD BOWL can serve as a cat's "hiding place," and you may find a toy or other small object neatly stowed there.

SPECIES WATCH

It is well known that gray squirrels bury nuts in the fall to eat through the winter. But a recent study has shown that they also make fake burials in order to fool any squirrel spies that might be watching. They ostentatiously dig a hole, and then fill it in again, while all the time the real stashes of food have been carefully hidden elsewhere.

Why does my kitten...
poop next to the litter box?

Q I have never had a cat before, and am struggling to litter-train my new kitten. I was told that she was litter-trained already, and she has certainly got the idea that she needs to poop in a particular place. The trouble is that the spot she likes is next to the litter box, not in it. A friend has told me that the quickest way to train a cat is to rub its nose in the mess, but I can't bring myself to do it. Is there another way to encourage her to use the box?

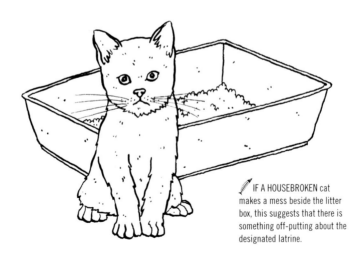

IF A HOUSEBROKEN cat makes a mess beside the litter box, this suggests that there is something off-putting about the designated latrine.

A A kitten's mother will usually show her how to use a litter box; most kittens will be using it by about five weeks. In theory, the kittens will take their good toileting habits to their new home, but as you have found, litter training isn't always that simple.

Your kitten knows her toileting spot, but finds the box off-putting for some reason. **Make sure that she can get into the box easily; an open box with low sides is best.** Your choice of litter is also important. Contact the original owner to find out what kitty litter your kitten is used to; you can switch to a different type later.

Site the box somewhere that feels safe to your kitten; the corner of a room is good. Do not put it next to the food bowl: No cat likes to soil its eating area. To start with, confine your kitten to the one room so that she can find the box easily. Place her on the litter on waking, after meals, and after playtime. Scratch the surface with her paw to get her used to the idea of digging. Clean the box regularly. Once your kitten is using the box, increase the area over which she is allowed to roam.

If you see your kitten in the act of pooping, quickly place her on the litter and praise her. This helps her to link the box with toileting. Never rub her nose in her poop: This will merely create anxiety in your kitten and make accidents more likely, not less so.

SHOWING YOUR CAT that feces belongs in the litter box can help it to learn to go in the correct place.

Why does my kitten ...
throw his toys up in the air when he is playing?

Q My kitten adores playing games, and will spend ages chasing and pouncing on a ball. I have noticed that after a while, he will slide his paw under the ball and then flick it right over his back. He then turns around and pounces on it all over again with renewed vigor. A friend told me that the kitten is pretending that he is stalking a bird—and that he flicks the ball up into the air to simulate a bird in flight. Is my kitten really that inventive? He's never been outside, so he hasn't been been anywhere near a real bird.

FLICKING A TOY up into the air is a common feature of cat play; it is a way of perfecting the hunting technique.

A This is a very common interpretation of normal cat behavior, but it is not quite right. The tossing action that you describe is not related to catching birds in flight; it is actually the action that a cat uses for fishing. An angling cat sits by the edge of the water; when a fish passes by, the cat slips its paw into the water and under the fish. It flicks the fish out of the water, over the cat's back, and on to dry land. The cat then turns and pounces on his prey to finish it off.

The fisher behavior is often misinterpreted because we are less likely to see our cats fish than stalk birds. A kitten that has free access to a pond or river will exhibit this behavior in real-life circumstances, and will be able to land a fish by the time it is about seven weeks old.

So your friend is correct when he says that, by batting and tossing toys about, your kitten is practicing ways to deal with prey. Stalking and pouncing are instinctive activities, but they are also skills that need to be learned early, before a cat has to fend for itself. But your cat is not really "inventing" games. **While cats are remarkably fun-loving, and spend long periods enjoying their toys, their play is always purposeful.** Flicking a toy in the air is a mixture of an instinctive action and an acquired ability. Your cat is refining talents that he would need in order to survive in the wild.

CATS ARE NATURALLY expert anglers, but most cats do not get the opportunity to fish for real.

SPECIES WATCH

Many wild mammals are accomplished anglers. The brown bears of north-eastern North America know that in the fall salmon swim upstream to spawn, and they make the most of the opportunity. The bears wait at waterfalls, where they pluck the fish out of the air, snatching them in their jaws, as the salmon attempt to leap into a higher pool.

Why does my kitten...
hide away from me?

Q My new kitten is a pretty little thing, but very shy. Since I brought her home a few days ago, she has hidden under the couch practically all the time. She comes out when I am not in the room, but whizzes back when I open the door. I have tried sticking my head down to her level and holding out my hand to pet her, but she just retreats farther under the couch. I enticed her out with a fishing-rod toy I have—but that didn't work for long. How can I let her know that I am her friend?

MOST KITTENS ARE timid when they are introduced to a new home, and may need time to get used to their human companion.

A It's quite normal for a kitten to hide away while it gets used to a new home. Kittens find this process easier if they have one of their siblings with them, which is why many people choose to adopt two cats from a litter rather than just one.

It is a good sign that your kitten has come out to play once or twice. But it will take time for her to get her confidence. You can help by confining her to one small room at first: This makes her new territory a manageable size. She should have a litter box, food and water bowls, and toys here.

Spend plenty of time in the room with her, but don't engage with her. Unlike humans, cats are most comfortable when they are being ignored. Once she understands that you are not trying to grab her, she will relax in your presence. You should also avoid staring at her, since she will perceive this as a threat.

IF YOU ARE patient and calm, a reserved kitten will slowly warm to you, and may eventually come to sit on your lap.

Talk softly and quietly to her whenever you feed her so that she gets used to your voice. Leaving the radio on will help accustom her to human voices generally. Carry on using the fishing-rod toy, but do it nonchalantly so that you do not frighten her off, and keep play sessions short. Try petting her gently from time to time, so that she gets used to your touch. Slowly increase the amount of time you spend in the room with her. In time, you are sure to find that she will initiate play with you, and will come to you for affection.

Why does my kitten ...
swipe at my older cat?

Q I have a much-loved fourteen-year-old cat, and have just adopted a stray kitten from a rescue home. I was concerned that my cat would bully the kitten, but it seems to be the other way around. She is constantly bothering him, and has often cuffed him with her paw. He seems to be responding by avoiding her, and has stopped coming into the living room or kitchen when she is there. They are indoor cats, so need to get along in the same space. How can I help them to establish friendly relations?

OLDER CATS ARE usually tolerant of a new kitten, but they are unlikely to enjoy the boisterousness of their young companion.

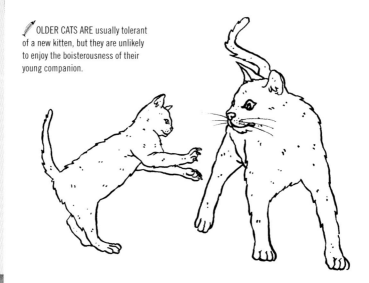

A Cats are naturally solitary animals, so introducing a new one into the household can be tricky. Your kitten is less of a threat to your older cat than another adult would be, but your tom clearly does not like having to accommodate a boisterous youngster. **Elderly cats can find it stressful to have a kitten around, and will retreat to a quiet spot in the house to avoid them.** Full-grown cats will not usually attack a kitten, though most are not above giving the youngster a corrective swipe. When your kitten strikes the older cat, on the other hand, she is probably inviting play.

The best way to help your tom cope is to get a second kitten: the young cats will play together and will leave your elderly tom in peace. If that isn't possible, then keep your new kitten confined to one area of the house, and bring the two cats together only in your presence while they get to know each other. Give the kitten plenty of toys, and remove her if she is bothering your older cat. Make sure you give your tom as much attention as before. This will reassure him that he is still the dominant cat in the household.

Also, make sure that there are enough litter boxes—indoor cats need one each plus an extra one—and provide separate feeding stations. As long as they do not have to fight for resources or for your attention, your cats should eventually learn to live together peaceably.

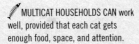
MULTICAT HOUSEHOLDS CAN work well, provided that each cat gets enough food, space, and attention.

Why does my kitten …
bite and suck on my shirt as I cuddle her?

Q I have had my kitten since she was six weeks old, and she is very affectionate. She jumps on to my lap and pads her little feet up and down on my chest as I stroke her. Somewhat less appealingly, she dribbles from time to time. She will also suck on my clothing and she will bite a little of the material and pull on it. My shirt can become really wet. Is there anything that I can do to stop her doing this?

MANY CATS WILL knead as you pet them. This behavior is a throwback to kittenhood, when they made the same action to stimulate their mother's milk flow.

A All cat owners will recognize the behavior that you describe. It is usually termed "kneading," and it is something that cats commonly do as they settle down on your lap. The only problem with kneading is that the cat exposes its claws while doing it, so it can be quite painful to be on the receiving end of a kneading if you are wearing thin clothing, or if your cat's claws aren't clipped.

Kneading is actually a throwback to kittenhood: It is something that kittens do to stimulate the mother's milk flow. (Human babies do a similar thing: they "bat" their mother's breast when they want to feed.) Sucking is a less common extension of this behavior. Your cat is replicating the suckling action she used at her mother's nipple. All cats see their owners as a mother substitute, and can revert to kittenlike behavior during petting sessions. But sucking is something that tends to be done by cats who have been weaned too early. You brought your cat home at the very young age of six weeks; a kitten is usually weaned by six or seven weeks, but should ideally remain with its mother until 12 weeks.

Sucking kittens can grow out of the behavior, but many don't. Some progress to sucking on your earlobe, and wool-eating (*see page 100*) is another related behavior. Many owners simply allow their cats to suck, but if you find it disturbing, try giving your kitten a soft pillow to use as an alternative. Whenever your cat starts to suckle, place her on the pillow. Introducing focused playtime may also be of benefit in encouraging a nervous cat to become more independent.

SUCKING ON SOFT material is a behavior typical of cats that have been separated from their mothers at a very young age.

Why does my kitten ...
keep getting stuck up a tree?

Q My poor little kitten loves climbing trees, but he seems much better at getting up them than getting down. The other day, he was stuck up a tree for six hours and I had to borrow my neighbor's ladder so that I could rescue him. I'm worried that one of these days he is going to go too high for me to reach him. How can I encourage him to stay on the ground?

CATS ARE NATURAL tree-climbers and use their tail to maintain their balance. Indoor cats will also seek out high-up places in the home.

A Tree climbing is a natural behavior for cats. In the wild, trees are the obvious place in which to hide from predators. So it's futile to try to stop cats from climbing if you let them outdoors.

But you are right to think that your cat is better at getting up than down. A cat's claws are curved in such a way as to grip a tree trunk on the ascent. **If a cat tries to come down head-first, it is sure to lose its footing. It can take cats some time to figure out that they have to come down butt-first.** That said, most cats do eventually make it back to the ground. As the saying goes: You never see a cat skeleton in a tree.

You may be assuming that your cat is stuck when actually he is perfectly happy. So, in the first instance, you should just wait. If hours have passed, you can encourage him by calling gently and calmly. Don't shout: You will scare him, which will make him climb higher. A tempting can of tuna fish at the foot of the tree can often provide a stuck cat with the extra courage he needs. It may also help to lean a wooden ladder against the tree, but your cat is more likely to use it if given privacy, so leave it in place for half an hour or so.

If none of this works, then you might resort to going up after him, or getting someone else to do so. If so, then make very certain that the person climbing the tree takes all necessary precautions: humans are, after all, far less agile and nimble than cats.

A CAT CAN climb up a tree more easily than it can come down again, but very few cats genuinely get "stuck."

Chapter two
A tiger in the home

The cat has adapted remarkably well to domestic life—so well, in fact, that it can live happily as a farm cat with acres of space, or as an indoor cat confined to a small apartment. **Part of the cat's enduring appeal is its clear pleasure in being petted. But there is another side to your cuddly** *Felis domesticus*: **It still bears the genes of a wild, lone hunter.** This section looks at the untamed essence of the domestic cat. It is this aspect of its heritage that explains why a cat may chatter its teeth when it looks out of the window at potential prey, how it broadcasts signals using its ears and tail, and why it may appear to grimace as it sniffs an unusual odor. Many of the behaviors described here are things that all owners will recognize. Knowing what lies behind them will give you a deep insight into your cat's special nature.

Why does my cat...
behave strangely before a storm?

Q Both my cat and I hate thunderstorms. She will hide under the bed while they are going on, and I must admit that I feel like joining her. Recently, I have realized that she behaves strangely before a storm, too. She will seem very nervous, and will sometimes disappear even before the thunder starts; when I go to find her, she invariably seems to be grooming herself. Can she predict the weather?

IT'S NOT UNCOMMON for a cat to hide before a thunderstorm breaks; some people say that it will groom the back of the ears.

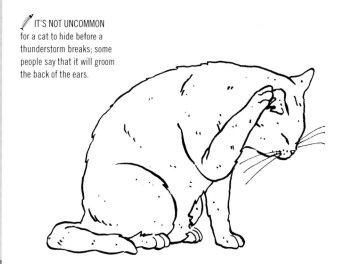

A There is an old superstition that if a cat washes behind its ears, then a storm is coming. Since cats groom all the time, this isn't really a useful way of predicting the weather. But cats and other animals do seem to sense when bad weather is brewing, possibly because they detect barometric changes in the air. Some humans claim the same ability: Among people who regularly suffer from migraine headaches, there are some who claim that attacks sometimes coincide with the onset of stormy weather.

Cats and other animals are also widely believed to sense the approach of rather more violent natural events, such as earthquakes. Many owners report that their animals behaved strangely or made themselves scarce before a quake. Most experts believe that animals do have some kind of "early warning system" in place, an inner seismograph that tells them to flee the danger.

It's not clear how this works, but it is possible that the animals pick up on the buildup of static electricity that occurs before an earthquake—and before a storm, too. Another idea is that animals notice the subtle variations in the earth's magnetic field that precede an earthquake. These heralds go unnoticed by humans; but animals, with their more finely attuned sensory array, read the signs and act accordingly. Yet another theory is that animals' refined sense of smell allows them to detect the subterranean gases that might escape through the minute fissures in the earth's surface when an earthquake is about to occur. So, your hunch that your cat senses an approaching storm has a whole raft of scientific theory behind it, and may well be true.

CATS AND OTHER animals are known to have fled places where an earthquake, tsunami, or other natural disaster was about to strike.

Why does my cat...
chatter his teeth
when he sees a bird?

Q My cat loves to sit at the window and watch the world go by. Not surprisingly, he pays great attention when a bird comes into view. But the weird thing is that he also makes a kind of chattering sound with his teeth. It is the oddest noise, and I have only ever heard him making it when he is watching a bird through the window. What on earth does it mean?

A WINDOWSILL IS a favored perch for many cats; they seek out high places where they can get a good view of prey.

A This is one of the more peculiar cat behaviors that you are likely to witness. **Chattering is part of your cat's hunting instincts, and he will only do it when he is observing prey;** some cats have been known to do it when they see birds or squirrels on television.

As you've found, it is the strangest of noises—a kind of stuttering rattle. Experts used to think the noise was an expression of extreme excitement or frustration: the result of the cat seeing a bird or animal that it could not reach. Another theory was that the sound was a threatening noise. This seems unlikely, since the chattering noise is bound to scare off prey, and cats are—sensibly enough—usually silent when stalking their prey. More fanciful still is the idea that the teeth-chattering is a cat's attempt either to replicate birdsong or to reproduce the sound of a bird's own prey, such as a cicada, with the intention of bringing the prey closer.

However, most experts now believe that the sound indicates that the cat is practicing the special bite it uses to kill its prey. In other words, your cat is anticipating the moment when he will have his victim pinned down by his claws and can dispatch him. Cats do this by biting down into the back of the prey's neck and then using a vibrating action to "saw" through the spinal cord. This special action brings about almost instantaneous death—and careful observation shows that cats perform the same jaw movement when they chatter.

A CAT THAT is about to pounce may sway its head from side to side to help it judge the distance accurately.

Why does my cat...
pull her food out
of the bowl?

Q I bought my new cat a beautiful bowl to eat from, but when I give her canned food, she insists on taking it out of the bowl and eating it off the floor. If I put out dry food, she sometimes leaves it scattered all around the bowl. What is going on? My cat is fastidiously clean the rest of the time, so why is she so messy when it comes to eating?

IF YOUR CAT eats off the floor, this may mean that the shape or size of its food bowl is unsuitable, or that the food is served in too large a lump.

CATS PREFER TO eat from a large, shallow bowl so that they can get at their food easily, without having their sensitive whiskers squashed.

A Your cat is not being deliberately untidy when she eats off the floor. She's just eating her food in the easiest, quickest way possible. Cat food often comes out of the can in a lump that is too big for her to manage. Pulling food on to the floor makes it easier for her to bite it into smaller chunks. Try cutting her food up before you give it to her; that may be all it takes to get her to eat from the bowl.

If that doesn't do the trick, you have probably bought the wrong sort of bowl. If it is too small or high-sided, then your cat's whiskers may touch the sides of it while she is eating. **A cat's whiskers are sophisticated motion-detectors, and cats dislike having them bent or squashed.** A large, shallow bowl may suit her better. Persian cats, with their flattened faces, also prefer a shallow bowl because they do not like having to push their entire face into the dish to get at the food.

As for scattering her food, your cat may simply be telling you that she has had enough. Cats prefer to eat little and often, as they would have done in the wild. One study found that cats given free access to food ate more than thirty times a day. Not many owners have the time to place a tiny portion of food in a freshly cleaned bowl that often, but you might want to consider giving your cat smaller, more frequent meals.

Why does my cat…
love catnip, when my other cat shows no interest in it?

Q I have two cats, a male and a female. I planted some catnip in my garden for them to enjoy, but have been surprised at how different their reactions have been. The boy just gave it a sniff and walked away, but the girl went crazy for it. She licked it, chewed it, then rolled all over it, leaped about, shook her head, growled, and generally indulged in some pretty frenzied behavior. She seemed perfectly fine afterward, and I have been told that catnip is quite safe. But why did something that had such a strong effect on one cat hold no interest for the other?

A CATNIP-INDUCED frenzy tends to last ten minutes or so before the cat returns to normal. There are no lasting effects from it.

A Watching your cat try catnip for the first time can be a interesting experience. As you saw with your female, a cat can engage in dramatic, almost "trippy" behavior. Some cats, on the other hand, react by becoming very relaxed and going into a trancelike state. Lions and other feline species react in the same way as domestic ones.

Catnip is not a drug, but a herb of the mint family. **The active ingredient is a substance called nepetalactone, which is released when the leaves are broken.** Scientists do not know why the substance affects cats so strongly, but the effect is short-term, and seems to cause no harm. So, it is safe for your cat to chew the leaves or play with catnip-stuffed toys. Very occasionally, a cat can eat too much and vomit, but this will resolve itself naturally.

But not all cats go catnip-crazy. The herb has no effect on kittens under the age of three months, and older cats also respond less to it than younger ones. Some cats, such as your male, do not react at all: It is estimated that up to a third of cats are unaffected by nepetalactone. This is down to genetics (your cat's parents would have been similarly unmoved), so it is nothing to worry about.

One word of warning, though. Like all mints, catnip is an invasive plant and can quickly grow out of control. Ideally, plant it in a large pot to confine the roots, and sink the pot into the flowerbed.

THE CAT WILL bite or chew catnip leaves in order to release the active ingredient, which is a chemical called nepetalactone.

Why does my cat...
curl his top lip when he encounters a strong smell?

Q A neighbor's cat got into my house and sprayed the couch while we were in the garden. My cat seemed horrified when he noticed. He went up to the sofa and smelled it very intently. Then he drew back, opened his mouth, and curled his top lip back, staring into space like he was hypnotized. I have never seen him grimace or look so spaced out before. Is it something that cats do when they are repelled? If so, why?

THE CURIOUS GRIMACE that a cat makes when encountering certain odors is known as the Flehmen response. You see it in other animals, too.

A When cats are disgusted by a smell, they draw back from it; they want to get away as quickly as possible. But your cat's response to another cat's spray was different. It was not revulsion; his reaction shows that he was deeply interested. Cats' spray contains a great deal of information, and your tom wanted to learn everything he could about the cat who had left a marker in his territory.

It is a fact that cats have a much more highly developed sense of smell than humans. It is their primary sense, in the way that sight is for us. It is not just that cats can detect smells more keenly. They actually have an additional scent organ in the roof of the mouth. It is called the vomero-nasal organ, or VMO. **This olfactory superpower allows a cat to "taste" an odor in a way that gives it lots of vitally interesting data about any smell.**

When your cat made the strange grimace that you describe, it was in fact activating the VMO. To do so, it must raise its chin, open its mouth, curl back its lip, then inhale slowly. The strange expression closes off the usual route of its breathing and directs the airflow over the VMO, allowing the cat to conduct a kind of chemical analysis of the odor. This takes concentration, which is why your cat seems to go into a trance. He is seeking an answer to the key question: was this marker left by a female in heat, or by a male muscling in on my patch?

A SENSE OF smell is vital to a cat. Your cat will always sniff its food before eating, in order to check that it is fresh.

Why does my cat...
lick his lips so often?

Q My cat has a funny little habit of flicking his tongue out to lick his lips. At first, I thought he was doing it to clean around his mouth, but he doesn't seem to do this at times when he might be dirty, such as after eating or when he is grooming himself. I have noticed that he tends to lick his lips when I switch the air-conditioning on. What does this mean?

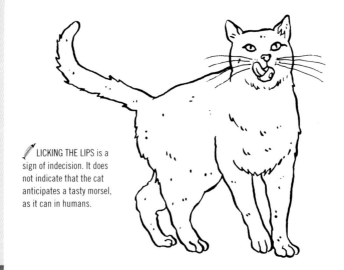

LICKING THE LIPS is a sign of indecision. It does not indicate that the cat anticipates a tasty morsel, as it can in humans.

A Cats, like many animals in the wild, have a two-way choice in dangerous or unpleasant situations: stand and fight, or turn tail and run away. But some situations are not clear-cut enough to trigger the fight-or-flight reflex, so the cat resorts to a third, noncommittal response. **This is what body-language experts have dubbed a "displacement activity," and it is common in humans, too.** When you are feeling uncertain, you might fiddle with your hair or bite your fingernails. Licking his lips is your cat's equivalent of these actions—a kind of behavioral tic to help him to expend the nervous energy that is building up.

A cat is likely to indulge in displacement activity when it is caught between two conflicting desires: for example, seeing something that catches its attention, but is at the same time confusing or distressing in some way—perhaps hearing a cat yowling on the television, for example. When your air-con begins to hum, the cat is not sure whether to get up and leave, or stay and investigate. The licking action is a way of displacing the agitation that he is feeling, because it is a comforting and familiar activity.

Grooming is another very common displacement activity. You'll often see a cat washing itself when you have visitors, for example. It's also normal for a cat to attend to its fur a lot more often if its home circumstances suddenly change. So grooming is something that a cat will do to help it to cope with the stress of, say, a house move or the arrival of a new baby in the family (*see also overgrooming, page* 104).

SPECIES WATCH

Many animals resort to displacement activity when stressed. Conflicted chimpanzees, like humans, have been known to scratch their chins. Some species of birds will peck at grass or rub their beaks against a branch. And if two male sticklebacks come into conflict but are not sure whether to fight, they point their heads down and dig, as if making a nest.

LIP-LICKING IS JUST one gesture that a cat makes when it is uncertain. Others include twitching the tail or full-scale grooming.

Why does my cat…
play with her prey?

Q My partner claims that cats have a vicious streak, and he cites the way that they play with their prey as proof of this. On the odd occasion that I have seen my cat catch a mouse, he has certainly seemed to enjoy tormenting it. He will pretend to lose interest, allow the poor mouse to make a run for it, and will then pounce again. Is he just doing this out of spite, or is there another reason why he toys with his unfortunate victim?

A DOMESTIC CAT may keep a mouse or other small prey alive simply so that it can prolong the game of hunt-and-capture.

A It can certainly be distressing to see your beloved cat seemingly torture another living creature. But rest assured, your cat is incapable of the kind of malice that your partner ascribes to him. One might think that this behavior is another example of an instinct that has survived domestication, that what you are seeing is "nature red in tooth and claw." But, strangely, this behavior is not found among wild species of felines, and only rarely among farm cats. Only household cats torment their prey in this way. It's a reflection of the fact that they do not have sufficient opportunity to hunt. **When a house cat does get its paws on live prey, it is naturally reluctant to end the fun too quickly.**

Another apparent "game" that a cat will play with its prey is striking it with its paws rather than dispatching it with a bite to the neck. In contrast to the capture-and-release game that you describe, there is a natural explanation for this. Some larger prey—such as a rat or a vole—is capable of giving a cat a nasty bite. By striking the animal repeatedly, the cat beats it into a daze. It can then dispatch its prey without risking injury. A cat that hunts a lot does not strike a mouse in the same way, because it knows that the mouse cannot cause it harm. A household cat, which has less experience of hunting, is unsure about the risk that a mouse poses, so it will beat it into submission to be on the safe side.

A KITTEN LEARNS to deal with prey by batting a small object around, and pouncing on it.

Why does my cat...
have "crazy" times?

Q Every so often, my cat will spend several minutes acting in a quite demented fashion. He will run round the house quite manically, he'll leap about and pounce on imaginary prey, and he'll chase his own tail. Sometimes he'll calm down and then have another bout of crazy behavior a bit later. It usually happens late in the evening. He has a really wild look about him while he does this, and I wondered if it was a sign of mental disturbance.

TAIL-CHASING IS A behavior more often seen in dogs, but cats can do it, too, during "crazy" times.

A Don't worry. Lots of people have reported their cats engaging in frenetic chases around the home for no apparent reason. It isn't exactly normal behavior, in the sense that you wouldn't see a cat behaving like this in the wild. However, there is a perfectly rational explanation for it.

It's probable that your cat, like most household cats, spends a lot of time indoors. While he is no doubt well fed and cared for, he may not be getting enough opportunity to run and chase prey (a cat can hit a speed of around 30 mph if given enough space). Inevitably, his natural instinct to hunt is somewhat thwarted. On a day-to-day basis, this isn't a problem. Cats are immensely adaptable animals and tend to be very happy living in households. But every now and then, the urge to give chase becomes overwhelming. Some small stimulus—which is often unnoticed by the owner—sets off that primal instinct, and the cat zooms around.

This doesn't mean that your cat is unhappy. **His crazy times are a strategy that he employs to release pent-up energy, and to help keep himself in shape.** However, if you want to be sure that your cat is getting enough stimulation, you could think about increasing the amount of time that he plays. The best way to do this is to give him some toys: try a paper bag, a scrunched-up paper ball, or a catnip mouse (available from pet stores); it is a good idea to put the toys away and then to reintroduce them from time to time to keep him interested. Your cat will also greatly appreciate a few minutes' dedicated playtime with you each day.

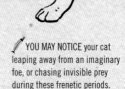

YOU MAY NOTICE your cat leaping away from an imaginary foe, or chasing invisible prey during these frenetic periods.

Why does my cat...
swish his tail when he isn't cross?

Q Dogs wag their tails when they are happy, and cats wag theirs when they are angry, right? That's what I thought before I got a cat, but now I am not so sure. My cat seems to swish his tail throughout the day. Sometimes I can see that he might be annoyed, but at other times he seems happy enough. So, does tail swishing have more than one meaning?

THE TAIL IS a useful barometer of a cat's mood. Here, the cat is signaling indecision—is it time to get up or return to sleep?

A If you want to understand your cat, his tail is a good place to start. A cat's tail is one of the main indicators of a cat's mood, but tail language is a lot more subtle than most people think. For instance, a swishing tail isn't always a sign of anger. More often than not, it is a sign of indecision or uncertainty.

Generally speaking, if a cat is thrashing its tail in broad sweeps, you can be pretty sure that it is annoyed; very fast, broad sweeps mean that it is about to attack. Cats will also swish their tails, or twitch just the tips, when they are curious or uncertain. If your cat wakes from a sleep in the sun, then spots something that interests him, you will notice that the tip of his tail starts to wag before he gets up.

Cats use their tails to communicate, too. For example, they raise their tails high as a greeting. It is thought that the raised tail serves as an invitation for tail twining and mutual rubbing. A cat's tail goes stiff and quivers when it greets its favorite person—generally its owner. When greeting a stranger or another cat, on the other hand, a cat's tail is often hooked, indicating that it is feeling friendly, but slightly wary.

Frightened cats, by contrast, tend to carry their tails low. If the tail is tucked between the cat's legs, then the cat is showing total submission. If the tail is fluffed out, then the cat is afraid. You will also see a bristled tail if the cat comes across an aggressor that it is trying to bluff into thinking that it is a force to be reckoned with (*see also page 48*).

SPECIES WATCH

Most animals would rather face down an attacker than fight, so attack signals are almost universal in nature. Skunks turn their head and aim their rear end toward an attacker as a sign that they are about to spray. Species as diverse as mandrills and zebras open their mouths and display their teeth to convey an idea of the weaponry at their disposal.

RELAXED CATS WALK with their tails down, curving up at the tip. The tail can quickly come up if the cat spots a friendly feline or human.

Why does my cat…
chase my neighbor's large dog?

Q My neighbor has a Doberman pinscher. The dog is usually kept in its own yard, but from time to time he gets free. My cat is one tough cookie, and manages to keep the other cats in the neighborhood in check, but imagine my astonishment when I saw her chasing the Doberman across a field. I have now witnessed my cat see off this dog on several occasions, and clearly neither she nor the dog have worked out that he is the stronger one. How does my cat manage to frighten an animal several times her size?

A CAT MAY bluff a dog into thinking that it is a formidable opponent, and may then press home the advantage by chasing the canine away.

A Clearly, your neighbor's dog is capable of beating a cat in a straight fight. What has happened here is that your cat has fooled the dog into believing that she is dangerous—and now she habitually presses home that psychological advantage.

Often, a cat will run away from a dog and take refuge in a high, safe place, such as a tree. **But a cornered cat has to find another way to deal with the aggressor.** In effect, it has to bluff its way out of trouble. So it puffs itself up in an attempt to look bigger: it arches its back, stretches its legs to their full lengths, and stands its fur on end. All this is an attempt to convince the dog that it is a worthy opponent. The cat will also stand sideways on to the dog, so that it is seen from its biggest angle. It hisses and spits, then lashes out with its claws. This display might well be enough to overawe a dog, and to force it to beat a retreat. Sometimes—as in your cat's case—the smaller animal then gives chase, knowing that it has its enemy at a disadvantage.

It's worth pointing out that dogs are most likely to fall for the cat's fight-to-the-death display only if the cat adopts this at the beginning of the encounter; as always in life, the bluff has to be wholehearted to be convincing. If the cat runs off in the first instance, that will trigger the dog's hunting instinct—and it will give chase. Once the dog has done this, it is unlikely to be fooled by a cat's display of bravado.

FACED WITH A bigger foe, a cat may lift its hair on end and stand sideways in order to increase its apparent size.

Why does my cat...
eat grass and houseplants?

Q My cat seems to have a liking for grass. Every now and then, he will chew on a clump of it and swallow a few blades. Most of the time he seems fine afterward, but sometimes he will regurgitate the grass, together with the rest of the contents of his stomach. Recently, he has also taken to chomping on some of my houseplants: I came back from work the other day to discover that one of them had been judiciously pruned. How can I stop him?

EATING GRASS MAY make a cat regurgitate, but there are no known harmful effects, and this behavior is thought to be beneficial.

A What you have observed is completely normal cat behavior—and something that is almost impossible to prevent. It seems strange only because we think of cats as carnivores. But the fact is: most of them eat a little vegetable matter from time to time.

Nobody knows exactly why cats do eat grass, but there are various theories. One is that cats eat plant matter to add roughage to their diet. But the amount that they digest is so small that it is hard to see how this could be the case. A more convincing theory is that cats use grass as an emetic, that is, eating grass causes them to vomit in order to help remove a hair ball or something else that disagrees with them. If your cat is doing this regularly, see a veterinarian to make sure that nothing is wrong.

Cats don't always vomit after eating plants, so there must be some other reason. The latest idea is that cats eat small amounts of grass to obtain folic acid. This essential B vitamin helps in the production of normal red blood cells, and so prevents anemia.

It's not surprising if cats who eat grass also nibble on houseplants; some even try cut flowers. But certain plants and flowers—including poinsettias, ivy, azaleas, sweet peas, delphiniums, and daffodils—are toxic to cats. Ask your veterinarian for a complete list of poisonous plants and flowers. You can discourage your cat from eating plants by dusting the leaves with cayenne pepper, putting orange peel in the soil as a deterrent, or by moving plants out of reach. If you have an indoor cat, you may want to grow a tray of grass for your cat to nibble on; he should then leave your houseplants alone.

CATS KEPT INDOORS may attempt to eat houseplants as a substitute for grass.

Why does my cat…
twitch when she sleeps?

Q My cat loves to nap on my lap, and I enjoy watching her twitch and wriggle as she sleeps. From time to time, she will growl and even hiss. And once, she thrashed her legs, yowled, and snarled— then woke with a start. I am convinced that she is dreaming, and that she can have nightmares as well. My wife insists that cats do not have the imagination to dream. Who is right?

CATS CAN SLEEP for 16 hours a day. Much of their sleep is light, meaning that they can be fully awake in an instant.

ELECTROENCEPHALOGRAMS (EEGS) have been used to monitor the brainwaves of sleeping cats. The results have shown that cats are in deep sleep—rapid eye movement (REM) sleep—for about a third of their sleeping time, and that they dream during these periods.

A Though your wife thinks your cat is incapable of dreaming, chances are that she is wrong. Evidence suggests that cats dream as vividly and regularly as humans. Studies of feline brain activity have established that cats have sleep cycles, just as we do, and that they alternate between light sleep and deep sleep. During deep sleep, a cat will make small movements of the type you describe: The paws fidget, the claws come in and out, the ears twitch, the whiskers move, and the cat sometimes makes noises. The fact that these movements coincide with periods of deep sleep means that the cat is dreaming, or doing something very like it.

Scientists can't tell us, of course, what the cat is dreaming about, but many owners believe that their cats dream of chasing prey or playing games. **And it is indeed likely that part of the function of dreams in cats, as in humans, is to process the experiences of the day**—the bad things, as well as the good. So cats probably do revisit unpleasant experiences in their sleep. Whether that means that cats have nightmares is harder to say, and is a question best left to philosophers.

One thing we know for sure about sleeping cats is that they have an uncanny ability to keep an ear open. If a cat hears a noise that signals danger or interest, it can be alert in an instant. So, a cat may happily sleep next to somebody using a hairdryer, but may wake up the moment you pick up the can opener.

Why does my cat...
sometimes fall on her side, not her feet?

Q I thought that cats always landed on all fours after a fall, but that doesn't hold true for my cat. A few weeks ago, I saw her lose her footing and fall off the windowsill. My heart was in my mouth, but, sure enough, she landed on all fours. But the other day, I tried to pick her up when she was lying on the back of the couch. For some reason, she struggled to get away from me, and ended up falling on to the floor below. She landed hard on her side, and winded herself. What went wrong?

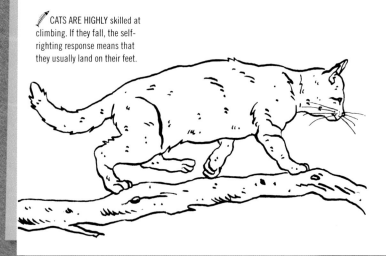

CATS ARE HIGHLY skilled at climbing. If they fall, the self-righting response means that they usually land on their feet.

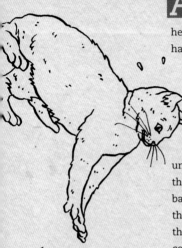

AN EXTREMELY FLEXIBLE spine allows the cat to twist its body mid-air so that it falls feet-first.

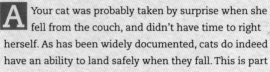

SPECIES WATCH

There are animals that know how to fall at least as elegantly as a self-righting cat. The flying squirrel has a sheetlike flap of skin, called the patagium, that connects its wrist to its ankle. When the patagium is extended, the squirrel can hang in the air like a kind of furry hang-glider, and can soar gracefully for distances of up to 65 yards.

A Your cat was probably taken by surprise when she fell from the couch, and didn't have time to right herself. As has been widely documented, cats do indeed have an ability to land safely when they fall. This is part of their natural armory of tools that has evolved to help cats cope with climbing. If a cat loses its balance, a "righting reflex" comes into play that enables it to adjust the position of its body as it falls.

Thanks to motion-capture videos, we now understand exactly how the righting reflex works. First, the falling cat establishes which way is up—using the balance mechanism of the inner ear. It turns its head the right way first, then twists the rest of its body into the "landing" position. It can perform this acrobatic contortion because it has an extremely flexible spine, and because it lacks a collarbone that would otherwise restrict its movement and slow it down. The tail also plays its part, acting as a counterbalance to keep the cat level as it falls. **At the last moment, a falling cat will arch its back and extend its legs toward the ground, which helps to reduce the force of the impact.**

But the righting reflex is not a guarantee against injury. In one study of cats that had survived falls from high-rise apartments in New York, many were found to have broken bones. Paradoxically, the farther a cat falls, the less likely it is to be killed or injured. This is because the "spread" position that a falling cat adopts slows its descent: The cat becomes, in effect, its own parachute. But it needs to fall for six or seven stories in order to relax into this position.

Why does my cat...
keep crouching down and yowling as if in pain?

Q My Oriental is five months old, and I am concerned about her. She keeps making this terrible screaming sound, as if she is in awful pain. She rolls on the floor and hunkers down on her front legs, with her rear end pushed up. I have also noticed that she is really clingy, and wants to rub up against me all the time. She did the same thing a couple of weeks ago and I was just about to take her to the veterinarian, when she seemed to get better. I have never had a cat before, and don't know what to do. Please help.

A QUEEN THAT is ready for mating will rest the front half of her body on the floor, and push the back half upward.

A Veterinarians often get calls from anxious cat owners who are convinced that their cat is in excruciating pain because she is behaving in this odd and distressing way. But there is a very simple explanation for it: Your cat is in heat. **All her behavior is designed to let the local toms know that she is fertile.** The call—which can be piercing, especially in the already vocal Orientals—is the most effective way that a queen can let potential partners in the local area know that she is fertile. She naturally takes up the mating position, pushing her rear into the air and sometimes moving her tail to one side. Many owners find that their cat is unusually affectionate when she is in "estrus" (the scientific term for in heat). All of the rubbing and rolling that you describe is her way of distributing her scent over as wide an area as possible, another way of conveying her readiness to mate.

Cats reach puberty between four and ten months. The average is six months, but Oriental breeds are often earlier. Usually the cat will come in heat during the summer months, when there is more daylight. She will remain so for five to seven days, and then will come in heat again every three to four weeks until the season ends, or she becomes pregnant. If you do not want your cat to have a litter, then you should have her spayed once she is six months. In the meantime, do not let her come into contact with toms whenever she is displaying the symptoms of estrus.

ROLLING AROUND AND being unusually affectionate are other signs that a cat is in heat.

SPECIES WATCH

Most female mammals have times when they are in heat. It is only during estrus that they are interested in mating or can become pregnant. In dogs, for example, estrus occurs every six to eight months and lasts two to four weeks. Humans, however, are highly unusual in that both males and females are happy to mate at any time, regardless of the reproductive cycle.

Why does my cat…
never meow?

Q I have a three-year-old cat who is unusually quiet. I got her from a rescue home when she was a year old, and I have never once heard her meow. She was put in the home when her owners moved overseas, and as far as I know, nothing terrible happened in her past to cause her to be mute. She interacts perfectly normally in all other respects: For instance, she sits happily purring on my lap while I pet her. But sometimes she'll look up at me and open her mouth as if she is going to meow, but no sound comes out. Why not?

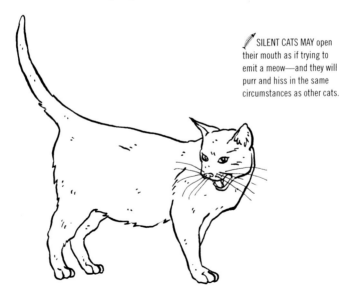

SILENT CATS MAY open their mouth as if trying to emit a meow—and they will purr and hiss in the same circumstances as other cats.

A It is by no means unknown for a cat to be silent; it is not even abnormal. Many cats go through life without feeling the need to meow; this doesn't signify that there is anything wrong or, as you have obviously wondered, that the cat has suffered a traumatic experience. You could even say that it is more normal for an adult cat to be quiet than otherwise, since in the wild, full-grown cats do not usually meow.

Domestic cats meow as a way of getting attention. It's a behavior that harks back to kittenhood. A young kitten that is away from the nest will often let out a series of distress mewls to help its mother locate it. You'll also find that a young kitten will mewl if you hold it, and quieten only once it is returned to its mother. House cats tend to retain this kittenish—one might say infantile—mode of behavior because they are fed and cared for throughout their lives, and never have to learn to fend for themselves. You could say that they meow because they have never grown out of the habit.

At the same time, it is certainly the case that some cats are more vocal than others. This is a matter of breed, as well as of individual personality. Siamese cats, and breeds derived from Siamese, have a persistent and distinctive meow that matches their highly demanding nature. The Russian blue, by contrast, is gentle, shy, and a cat of very few meows.

YOUR CAT MAY not meow, but will find other ways to communicate and make its wishes known to you.

SPECIES WATCH

Perhaps the most vocal species in the animal kingdom is the aptly named red howler monkey, which is native to South and Central America. Each day at dawn, troops of these surly monkeys set up a deafening chorus of hoarse whooping. This wake-up call is performed to stake out territory, and it can easily be heard three miles away across the dense forest.

Why does my cat…
change the position of his ears so often?

Q Why are cat ears so mobile? Most of the time, my cat's ears are upright, but slightly turned outward, but when he is really interested in something, his ears prick up. And once or twice, I have seen him flatten his ears completely—I think this was when he was frightened because there was another cat in the vicinity both times. Do cats have a special ear language?

A CAT WILL prick up its ears when listening intently to something, and may swivel one or both ears in the direction of the sound.

A As you rightly suspect, the ears are one of the whole battery of tools that a cat uses to communicate. We humans can be rather dense when it comes to the subtle range of body language that a cat employs. Learning a little about feline body language—and the part that ears play in it—is a great way of understanding your cat a little better.

A cat's ears are indeed extremely flexible: they can move up and down, and turn through 180 degrees.

The standard position for a cat's ears is pointing forward and slightly outward. This indicates a relaxed cat that is listening to the sounds around it. The ears become very straight (pricked) when the cat picks up on an interesting noise, and it may swivel one or both ears toward the noise. The tips of the ears may twitch if the cat is nervous.

In a conflict situation, a quick look at the ears of the cats involved will tell you exactly who is the aggressor. A frightened cat will fold its ears flat against the head. This is a defensive mechanism designed to protect the ears from injury during a fight. By contrast, a cat that is aggressive will swivel its ears sideways, so that the backs are visible. This is a sign that it is ready to attack—but this position also allows the aggressor cat to flatten its ears quickly if it is unable to intimidate the opponent and a fight ensues. Some large cat species have markings on the backs of the ears that serve to make the signals more visible.

THE SCOTTISH FOLD breed has permanently folded ears, which lends the cat a sweet and defenseless appearance.

Why does my cat...
know how to open doors?

I have always thought that cats were smart—they get us humans to run around after them, after all. But my latest cat, a Tonkinese, has learned how to pull the refrigerator door open with his paw, and he can open doors by jumping up at the handles, too. I haven't trained him to do it, so he figured it out all by himself. Is this a sign of intelligence?

SOME CATS DO not stop at opening the door of the refrigerator: They even help themselves to food from the shelves.

A Animal intelligence is a fruitful field of scientific testing. In the past, cats have often been considered less intelligent than dogs because they perform poorly in laboratory tests and can do fewer "tricks." In effect, cats were judged unintelligent because they didn't do what people wanted them to. These days, intelligence tends to be defined differently: It is held to be an animal's ability to solve problems that bring benefits or fulfill its own aims. This newer approach to animal intelligence has allowed psychologists to demonstrate what many cat owners have always known: Cats have their own way of being smart.

The fact that your cat can perform an apparently unnatural act, such as opening a door, is a sign that he can think on some kind of rational level. This is unarguably a form of intelligence. Cats figure things out by trying various possibilities until they hit on something that works; that is how your cat learned to open doors. Cats also have a good long-term memory (much better than that of dogs) which helps them to solve new problems, such as, say, finding a route to a high window.

A cat's curiosity and desire for stimulation can be harnessed to train it. Many

cats will learn to play fetch or walk on a leash, for example. But cats are easily bored, so training sessions will invariably be short. They will also only learn from someone they trust—another reason why they don't tend to do well in laboratory tests. Some breeds seem to be cleverer than others: Tonks, like Siamese and the Devon rex, are said to be among the most intelligent of breeds.

SPECIES WATCH

Dolphins are perhaps the most intelligent creatures in the animal kingdom. Wild dolphins seem to have a creative urge: They blow rings of bubbles, then watch—in apparent admiration—as their "artworks" rise to the surface. Dolphins have also been observed using "tools", tearing off a piece of sea sponge and using it to protect their noses while probing the abrasive seabed for food.

BENGALS AND SIAMESE cats are two feline breeds that like to carry objects in their mouths. This makes it relatively easy to train them to play "fetch."

Why does my cat...
sometimes purr when she is not happy?

Q One thing I thought I knew about cats is that they purr when they are contented. Sure enough, my lovely cat purrs away like an engine when I am petting her. But the other day, she had a horrid fright when some workmen fired up a pneumatic drill just near the bush where she was napping. Not surprisingly, she made her escape and hid under the bed, but, strangely, she also started to purr. She was clearly still distressed because I couldn't coax her out for more than an hour. So, why was she purring?

INJURED CATS ARE known to purr. This could be a way to calm themselves, but some scientists believe that the very act of purring can promote healing.

A You have raised a question that has perplexed scientists for years: Why do cats purr? Cats generally purr when they are being petted or fed, so clearly purring is sometimes a mark of contentment. Cats learn to purr when they are week-old kittens as a way of letting the mother know that all is well. She purrs in turn to let them know that she is relaxed and receptive. So purring begins as a form of communication. It is rooted in the relationship between a mother cat and her kittens, and so it is one of the signs that your cat considers you to be its parent.

MOTHER CATS PURR when nursing, to let their kittens know that all is well. They also purr when they are in labor.

Cats are also known to purr when they are stressed. For example, some cats purr when they are visiting the veterinarian, queens commonly purr when they are in labor, and cats have even been known to purr when severely injured, and even in the moments before death. The explanation for this paradox may be that the act of purring has a soothing effect on a cat. In the absence of another (mother) cat, a distressed cat may resort to purring in order to calm itself.

But some scientists believe that there may be another explanation. Cats purr in a consistent sonic pattern at a frequency of 25–150 hertz. This range of sound frequency has been found to improve bone density and encourage self-healing. So, the injured cat may instinctively purr in order to heal itself—surely a sign that the cat is a most remarkable animal.

Chapter three
You and your cat

The relationship between cats and humans is symbiotic, meaning that there is mutual benefit, albeit in different ways. Cats can provide their owners with a profound sense of companionship and comfort; in return, humans offer their cats a carefree existence that contains almost none of the perils and vicissitudes of life in the wild. This section focuses on the relationship between human and cat, **and looks at ways of establishing a close and respectful bond with this most independent of creatures.** The issues raised will help you to understand what your cat is doing when it brings a dead mouse into the home, or why it may suddenly bite you during a petting session. You may be surprised to learn how clever and adaptable a cat can be—and how often it turns out that the cat is training the human, rather than the other way around.

Why did my cat...
attack me when I went to calm her?

Q My cat has always been gentle and loving, but the other day she went for me. She was looking out of the window when a dog jumped over the fence into our garden. She arched her back, fluffed up her fur, and hissed. So, naturally I went over to calm her. I put what I thought was a reassuring hand on her back, but, to my horror, she lunged at me. It was a really scary experience. Why did she hurt me when I was only trying to help?

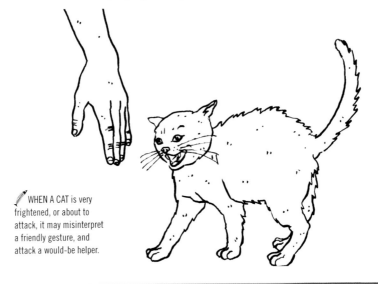

WHEN A CAT is very frightened, or about to attack, it may misinterpret a friendly gesture, and attack a would-be helper.

A You are describing a perfect example of how a well-meaning human can give a cat the opposite of what it needs, simply by misunderstanding feline signals. In this situation, your cat was showing all the signs of feeling terrified and cornered. You realized that she was scared, and went to offer reassurance—just what a child would need in similar circumstances. **But your cat is not a child; in danger situations, she reverts to being a wild creature.** When you touched her, she redirected all of her rising fear and aggression toward you. In that instant, she followed her deepest instincts and went for the nearest target as the most likely source of the threat.

This type of "redirected aggression," as it is known, is quite common. It is the reason why you should approach a scared or aggressive cat with caution. The object of the aggression need not be the cat's owner: It could be another cat or other animal, a human bystander, or a child. Sometimes, too, the cat can continue to attack the target of its aggression even after the danger has gone. That person or animal can become inextricably linked with danger or pain in its mind. In extreme instances, friendly relations cannot be re-established, and the cat has to be rehomed. Fortunately, that does not seem to have happened in your case.

A CAT CAN easily misidentify the cause of pain. If a human accidentally stands on its tail, it may attack a nearby cat.

Why does my cat...
sulk when I scold him?

Q My cat can be really cranky. Most of the time, he is very friendly and loving, but whenever I tell him off, he goes into a sulk. He'll turn his back on me, and he won't turn around if I call his name. He can definitely hear me (he puts his ears back when I am calling), so is he trying to get back at me, or does it genuinely take time for him to get over any upset?

CATS WILL COMMONLY turn their back on their owner after a scolding. This is a way of showing that they accept the rebuke.

A Your description of your "sulky" cat is one that many owners would recognize. But your analysis of his body language is wrong. You are assuming that your cat is behaving like a grumpy teenager, and are interpreting his actions accordingly. After all, when a person fails to answer you, this is often a coded way of saying that he or she is angry with you; but when a cat avoids eye contact, it conveys a very different message.

You need to understand the role of staring in the cat world. **To a cat, a direct, unblinking gaze is used to challenge or intimidate.** If a cat responds to another cat's hostile stare in kind, then a fight may result. If, on the other hand, it drops its gaze and looks away, it is signaling submission. By studiously avoiding eye contact, the cat is able to remain where it is, rather than retreating or having to fight.

When you rebuke your cat, you will almost certainly be glaring at him. You are a lot bigger than your cat, so he withdraws from this hostile situation and turns away from you. He is actually acknowledging that you are the dominant party, accepting your reprimand, and showing that he does not wish to defy you. If you want to put this into human terms, his actions are closer to an apology than a sulk. After a short period, your cat will usually relax and start to purr; he'll then be ready to resume friendly relations.

WHEN ONE CAT looks away from another, it is acknowledging the dominance of its rival, whose steady stare is a mark of superiority.

Why does my cat...
stick his rear end in the air when I stroke him?

Q I love cats, but there is one aspect of their behavior that has always puzzled me—and, if I am honest, troubles me. Why do they insist on sticking their rear ends in your face when you stroke them? It seems impossible to get a cat to sit on your lap without it doing this, and I find this behavior most off-putting. What is the explanation for it?

WHEN YOU PET your cat, you may find that it raises its tail to expose its behind. This disconcerting reaction is actually a sign of affection.

A This is a classic case of a communication clash between species. You, as a human, may find the sight of a cat's rear end rather distasteful; it looks to you as if your cat is somehow showing contempt for you. But in feline terms, what you are being offered is a very clear sign of affection. More than that, the display of the rear end is actually a signal that your cat has adopted you as a mother figure.

Cats enjoy being petted because it reminds them of the attention that they received from their mother as kittens.
From the first days of life, the mother cat uses her rough tongue to tend to the kitten's fur. She also licks its anus to encourage the opening of the bowels. When you stroke your cat with your hand, the sensation, from his point of view, is similar to a mother's licking. The fact that you pet your cat, and also feed and care for him, leads him to think of himself as your kitten. So, for him, it is natural to respond to your stroking as he did to his mother's touch—in this case, by showing you his anus. You'll notice that his tail is held stiffly upright at the same time; this instinctive posture has evolved to make absolutely certain that you get a clear view.

This is a natural behavior, which means that there is not a lot that you can do to change it. But understanding why your cat does it may help you to tolerate it more easily, at least.

THE DISPLAY OF the rear end is a hangover from kittenhood. When grown cats do it, it is because they see their owner as a mother figure.

Why does my cat…
spend so much time away from me?

Q My cat is spending less and less time with me. He is out during the day when I am home, and I barely see him in the evenings. It was never this way when I worked from home, but now I have an office job and am out early in the morning and home quite late in the evening. I would have thought that my cat would be extra pleased to see me when I get back, but he barely acknowledges my existence. Sometimes he misses meals, and once or twice he has stayed away for a couple of days at a time. Why doesn't he want to be with me any more?

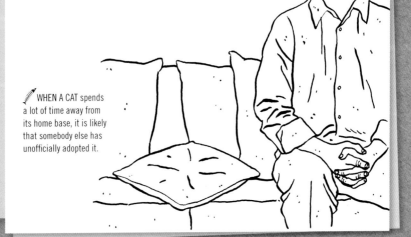

WHEN A CAT spends a lot of time away from its home base, it is likely that somebody else has unofficially adopted it.

A Cats are independent animals, but they do build bonds with their human companions. They are also creatures of habit and dislike changes to their routine. Your cat doesn't know that you need to work; all he understands is that his attentive companion is no longer available to him. You could say that he feels abandoned. There are countless stories of animals refusing to eat after their owner has died, and many people report that their cats are either standoffish or unusually clingy when they return from a vacation. **So, it is clear that a cat can be affected by its human's comings and goings.**

The fact that your cat's absences sometimes involve missing mealtimes is significant. No cat chooses to go hungry, so this suggests that your tom is being fed by someone else. This well-meaning neighbor may also be lavishing attention on your cat, allowing him into their home for longer and longer periods, including nighttime. It's possible that this person believes your cat is a stray or uncared for.

So talk to your neighbors to see if anyone is feeding your cat, and ask them to stop. Next, set regular mealtimes for your tom: This will help him to see your house as his home. Give him food that he likes to eat, and get an automatic feeder that will dispense food when you are not able to be home on time. Be available to your cat when you are at home, and invest in some toys and other devices, such as a cat gym, that you can tempt him with. If you allow your cat access to the outdoors, then it is best to keep him in at night, for safety reasons.

IF A CAT can choose to come and go, then there is always a chance that it may absent itself for long periods.

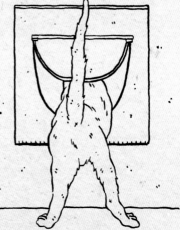

Why does my cat...
respond when I talk to her?

Q My cat definitely "talks" to me. When I come home from work, I greet her and she lets out a little meow in return. I ask her about her day, and she will respond to every question that I ask. She will keep up the conversation until I stop asking her things. I am quite sure that she understands every word I say, but my friends say that this is impossible. Who is right?

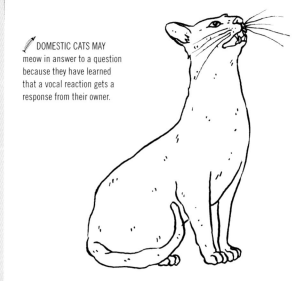

DOMESTIC CATS MAY meow in answer to a question because they have learned that a vocal reaction gets a response from their owner.

A Lots of owners are adamant that their cat understands them, and a few insist that their cat can actually say a few words. But there is a world of difference between communication and speech.

Cats communicate all the time, both with other cats and with humans. They use body language, facial expression, scent, and, to a limited extent, sound. **Domestic cats use sound with humans more than with other cats, because they have learned that this is a way of getting what they want.** In a sense, cats have cleverly adapted to our preferred method of communication, because we so often fail to understand theirs. In 1944, a researcher named Mildred Moelk identified 16 sounds used by cats, each of which, she said, had a distinct meaning. Behaviorists and many owners also testify that cats use different sounds, depending on whether they are trying to convey "I am hungry" or "I want to go out." All these sounds have developed from the simple cry for help that a kitten makes when it needs its mother.

So, it is certainly true that your cat voices her needs. It is also clear that cats do respond when they are spoken to, in the way that you describe. It is significant that cats owned by women are more likely to vocalize in this way; this is because female owners tend to talk to their cats. I'm afraid that your cat is not telling you what she did while you were away, but she is enjoying the greeting ritual that you have developed together as part of your special bond.

SPECIES WATCH

No animal has ever been shown to have a command of language. Even the well-publicized cases of captive chimpanzees using sign language are, at best, inconclusive. A dog responds to spoken commands, but that is not the same as having an understanding of speech. And the coherent language spoken by trained parrots is an illusion of speech, because the birds cannot use the words that they have learned to generate original sentences.

THE RAPT ATTENTION with which your cat listens to you is one of the things that makes it an appealing companion.

Why does my cat...
spend so much time grooming, when she is perfectly clean?

Q My female cat spends hours each day grooming herself. She's obviously clean, so why does she keep doing it? My previous cat—a tom—spent far less time tending to his fur. I wonder if there is a difference between the sexes here, but a friend says that her tom is a real groomer too. What's more, my cat seems to follow the same routine of wash and brush-up each time: She will wash her face, then work her way down the body, and finish up with her tail. Do all cats groom themselves in this systematic way?

CATS GENERALLY GROOM themselves from top to bottom, starting with the face and ending with the tip of the tail.

A Cats are famous for being fastidious about cleanliness, but individual cats vary in the length of time that they devote to grooming, as you know from your two. Most healthy cats will groom several times a day, and it is quite normal for a cat to spend a third or more of its waking hours on washing itself. Some cats are more interested in grooming than others, but this disparity has nothing to do with the differences between the sexes; an average tom is likely to spend as much time grooming as a queen.

A cat's tongue is covered with spiny hairs, which operate like the teeth of a comb. Cats are remarkably supple, and can twist to reach areas such as the middle of the spine. They will use the sides of their paws and forelegs to wash hard-to-reach spots, like the back of the ears. **And, as you have noticed, a cat will often follow a fixed sequence of grooming.** This regular approach simply helps to ensure that each part of the body is cleaned.

Grooming has functions over and above washing the fur and removing dead hair and debris. You will have seen that your cat doesn't just lick her fur; she will stop and tug at it from time to time. In this way, she stimulates special glands that release an oily substance to help keep the fur waterproofed. Grooming also plays a part in temperature control. Fur that has been licked smooth is a better insulator in cold weather. And in hot weather, grooming enables the cat to cover its fur with saliva, which then cools the cat's body as it evaporates. Cats need to do this, because, unlike humans, they have very few sweat glands.

SPECIES WATCH

Grooming is common to many animals, even insects. Ants smear themselves with an antibiotic produced by their own bodies—not so much to protect themselves as to prevent bacteria from taking hold inside the nest. In bees' nests, there are worker bees whose function is to groom the queen, then to distribute her scent among the colony: a kind of perfumed message that all is well with the hive.

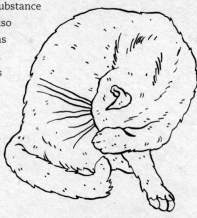

ITS TONGUE IS a cat's main grooming tool, but it will use its paws in places that the tongue cannot reach.

Why does my cat…
always plonk himself on my newspaper when I am trying to read it?

Q My cat tries to get between me and whatever I am trying to do. She loves to sit on the newspaper when I am trying to read it. She also seems to want to stop me using my computer; sometimes she will even sit on the keyboard in an attempt to stop me typing. I have also noticed that whenever I am on the phone, she jumps up on to my lap and pushes her face against mine, as if to move me away from the receiver. Why does she do this? I work from home, and love her company, but sometimes I find that the only solution is to shut her out of my office.

WHEN YOUR CAT jumps on to your newspaper, it is trying to get your attention rather than to prevent you reading.

A CAT RESPONDS to your voice, so may jump up to greet you when you are on the telephone.

A Cats seem to have an unerring instinct for making their presence felt. Sometimes it does seem as if they want to prevent their human companions from doing things that don't involve them. But they are not, as you might think, jealous of your newspaper or trying to annoy you. *Felis catus*, unlike *Homo sapiens*, takes no pleasure in riling other creatures.

It is true, however, that your cat notices when your mind is focused on something other than her. But when she jumps up on to your newspaper, it is not to stop you from reading it, but to get your attention. If you stop what you are doing to pet or play with her, she'll wander off in her own time. More prosaically, cats are also attracted to warm and comfortable places: the thick, dry surface of your newspaper, or the electric warmth of your computer keyboard, make for good places to sit.

Your cat doesn't know how a phone works, so when she pushes you away from the receiver she is not trying to stop you from chatting. The fact is: she hears your voice and assumes that you are talking to her; quite logically, from the cat's viewpoint, she comes to greet you in response. **Some owners report that their cat is more likely to interrupt a phone call if they are talking to a friend than if they are speaking to a work contact.** This makes sense, because your voice is likely to be softer and more inviting when you are speaking with a loved one than with an acquaintance.

Why does my cat... rub himself against my legs?

Q When I get home from work, my cat rushes to greet me. He presses his face against my legs, and against my hand when I reach down to stroke him. He'll do this for quite a while, and will push the side of his mouth or the top of his head quite firmly against whatever part of me he can reach. He'll also rub his entire flank along my leg, and will wind his tail around me. Is this just a feline version of a loving cuddle, or is there another reason why he does this?

RUBBING ALLOWS YOUR cat to leave its scent on you. Cats have scent glands on the face, the top of the head, and the base of the tail.

A Most cat owners enjoy the friendly welcome that they get from their feline companions when they get home. This affectionate behavior is often cited as evidence that the independent and aloof cat can make a loyal and loving pet. As is so often the case with your cat's behavior, though, it is not quite as simple as that.

Cats have special scent glands at the edges of the mouth and on the temples; there's also one at the base of the tail. When your cat rubs his head and tail against you, he is marking you with his scent (a substance called a pheromone). He also picks up your scent—this is the flank-rubbing part of the greeting. A cat has a highly sensitive nose, and it is important that his companions smell familiar. So, this is a ritual usually reserved for members of the household or other friendly cats.

When cats greet each other, they rub faces to share scents. You are too tall for your cat to reach your face, so he rubs against your legs instead. But because your cat's instinct is to rub faces, you may see him go up on to his hind legs when he first sees you—this is his attempt to get closer to your face. Some cats will jump up on to a high piece of furniture just so that they can rub cheeks with you.

If you observe your cat just after he has greeted you, you'll notice that he sits down and grooms himself. What your cat is doing is tasting the scent that he has picked up from you.

CATS WANT TO rub heads with you, which is what they do with other cats. When they jump up in greeting, they are trying to reach your face.

Why does my cat...
know when I am on
my way home?

Q Settle an argument for me. I am quite convinced that my cat is psychic; my husband says that I am being ridiculous. The thing is, my cat always knows when I am on my way home. Whatever time I get back—and it varies from day to day—he is sitting in the hallway, waiting for me. Is it possible that this is some kind of extrasensory perception, or is there a "rational" explanation for this?

CATS SOMETIMES SENSE when their owner is on the way home—not because of any feline extrasensory perception (ESP), but because they are aware of your daily routine.

A Cats certainly have an otherworldly air about them. It's part of their appeal, and it is the reason why, for centuries, cats have been linked to legends and tales of the supernatural. **Many pet owners believe, like you, that their animals can sense when they are on their way home.** However, this apparently prescient behavior is probably easily explained. To take your example, you may come home at different times, but if you drive home, your cat may well recognize the sound of your engine from afar (at least from farther than a human could), the sound of your garage door opening, or your footsteps on the pathway. He recognizes the signals, and so is always there to greet you.

Other stories about cats' psychic powers can be explained in the same way. For example, it was widely reported that a cat who lived in a nursing home in Rhode Island could predict when patients were about to die. The cat—Oscar—often curled up next to a patient who would die within a few hours. He did this so often (at least 25 times) that the staff would tell family members that the end was near if Oscar showed a particular interest in a patient.

Although Oscar made headlines as a psychic cat, most experts agreed that there was a biochemical explanation for his behavior. A dying person produces a particular smell that goes unnoticed by humans, but is picked up by the hypersensitive olfactory organs of a cat. Something about that smell drew Oscar to the dying patients. It was not a supernatural psychic phenomenon, but an entirely natural feline one.

SPECIES WATCH

Many animals have sensory abilities that, from a human perspective, are practically superpowers. Bats navigate in total darkness using "radar"; they emit high-pitched squeaks, then use the echo to make a mental map of the terrain. Many invertebrates see parts of the electromagnetic spectrum that are invisible to humans, such as ultraviolet light. And snakes detect predators and prey by feeling the vibrations of their footfalls.

CATS HAVE THEIR own ways of gathering information. Their hearing, sight, and sense of smell are all different from ours.

Why does my cat...
keep leaving bits of dead mouse for me to find?

Q My cat is a prolific hunter, and I have seen her dispatch various birds and small animals with great efficiency (only once have I managed to rescue an unfortunate mouse). I find this very difficult, especially as I am a vegetarian. Now, she is leaving body parts for me to find. I stepped out of bed this morning to find a mouse head and some innards on the floor. This repulses me. How do I stop my cat from doing it?

WHEN YOUR CAT deposits a dead mouse at your feet, it is trying to let you know that it is looking out for you.

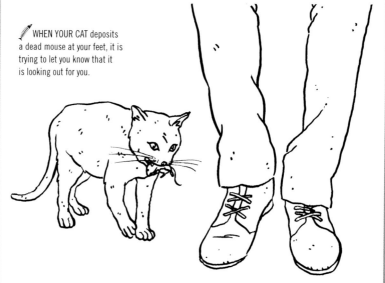

A Many owners find this a very difficult aspect of owning a cat. But you have to put your own feelings to one side and think like your cat for a moment. To her, killing prey is a natural instinct; after all, she wouldn't survive in the wild if she couldn't do it. She isn't killing because she is hungry; most household cats hunt for the excitement, rather than for food. Unless you keep your cat indoors all the time, you cannot stop her from killing.

The thing is: By bringing dead prey into the home, your cat is actually trying to please you. A mother cat will bring prey for her kittens to eat in the nest. In your cat's eyes, you are a very poor hunter—like a kind of overgrown kitten—who needs her help. The dead animal parts are, in other words, a caring gift. Interestingly, the cats who are most likely to leave dead prey for their human companions are neutered females; it is thought that they are channeling their maternal instincts toward their owners.

Giving your cat more food won't help; nor will scolding her. She won't connect sharp words or any form of rebuke with the dead mouse at your feet. She will merely be confused and afraid. So this is one situation when you will have to rein in your squeamish human sensibilities. Get rid of the carcass out of your cat's sight, and give her some affection and praise for her efforts: after all, it is the thought that counts.

SPECIES WATCH

Cats are not the only creatures that kill without eating. "Surplus killing" is a rare, but well-documented, phenomenon among wild predators, such as foxes and wolves. Their victims are usually captive farm animals, such as chickens or sheep. Perpetrators of surplus killing are not doing it for the sake of it. Most likely, wolves and foxes that are presented with an overabundance of prey simply cannot turn off their killing instinct.

IT IS IMPORTANT that you, as an owner, overcome any feelings of revulsion at your cat's kill, and praise it for what it has done.

Why does my cat...
head straight for visitors who don't like cats?

Q I know that some cats can be unsociable, but why do they tend to befriend people who can't stand them? Whenever I have a group of friends visiting, my cat will come into the room, walk past all the people who want to pet her, and rub up against the one person who dislikes cats. She also seems happier on my partner's lap rather than mine in the evenings, even though I am much more of a cat-lover than he is. Why doesn't she want to be with people who want to be with her?

CATS CAN MISINTERPRET the signals that humans give out. To a cat, an averted gaze means an absence of threat, not a lack of interest.

A A cat's tendency to gravitate toward people who are nervous of cats is sometimes seen as proof that it is an animal with a malign streak—or perhaps one with a perverse sense of humor. But this is to misunderstand how cats operate. What you are describing shows just how differently the human and the feline interpret friendly behavior.

When we greet somebody, we tend to look the person straight in the eye and smile. Normally, the person responds in kind; if not, we feel an instant sense of discomfort. **In the cat world, on the other hand, direct eye contact is used as a challenge.** If you watch two cats jostling for position, you'll notice that the dominant cat maintains an intense stare until the weaker one looks away, signaling submission.

If your cat comes into a room full of people, she will look for somewhere quiet to place herself. But she will quickly observe that most of the people—the cat-lovers—are staring at her. Perhaps some are calling her name, clicking their fingers at her, and so on. All of this is most off-putting to the cat. Then she notices someone who is ignoring her. She doesn't know that this person is trying to avoid her; all she sees is a display of non-confrontational behavior. To your cat, it makes perfect sense to gravitate toward this person.

As for her preference for your partner over you, there is probably a similar explanation. You may consider yourself the cat-lover, but ask yourself who is most likely to sit still and let the cat decide when to be friendly? The chances are that it is your partner, and that is why he makes a better pillow.

WHEN A HUMAN tries to make friends with a cat by petting it, this action can seem intimidating or off-putting to the animal.

Why does my cat…
roll on to her back even though she hates having her belly rubbed?

Q If I go into a room when my cat is sunning herself, she doesn't come to greet me. Instead, she rolls on to her back and stretches out her legs. To me, it looks as if she wants her belly stroking. But if I try, she swipes at me, and once she gave me a nip. Is there a right way to stroke a cat's belly, or should I leave well alone?

CATS ARE NOT like dogs: When your cat displays its tummy, it is not asking for a rub, but is simply greeting you in a deferential way.

A Wakeful cats tend to greet you by rubbing themselves up against your legs (*see page 82*). But a slumbering cat doesn't quite want to interrupt his sleepy mood by jumping up, any more than most humans would. Rolling over and showing the belly is the cat's lazy alternative—the equivalent of a human's shouted "hello" from the couch.

You may notice that your cat keeps looking at you while she rolls on to her back. This is because she is adopting a highly vulnerable position by exposing her soft underparts. The tip of your cat's tail will also probably be twitching slightly—a sign either of slight concern at her vulnerability, or perhaps the result of her being caught between the desire to get up and greet you and the wish to return to sleep. Either way, you can take it as a compliment that she trusts you enough to expose her belly; she will do this only to humans that she knows well.

But exposing her belly in greeting isn't the same as inviting you to stroke it. **Some cats do like having their belly rubbed, but many don't.** The belly is sensitive and vulnerable to injury, so your cat's natural instinct is to be highly protective of it—hence the swipe or nip when you go to stroke it. If a cat does let a human rub its belly, then it is a sign of an extremely intimate relationship between cat and owner. Even then, a cat is unlikely to put up with this for long.

SPECIES WATCH

Submissive signals are a vital social tool among animals that live in hierarchical groups or clans. Not surprisingly, some of the most sophisticated signals are to be found among monkeys. The stump-tailed macaque, for example, can signal submission to a more dominant monkey by presenting its hindquarters, smacking its lips, baring its teeth, and—most abjectly of all—offering its arm to be bitten.

YOUR CAT IS more likely to greet you from afar when it is feeling sleepy, or when it has only just woken up.

Why does my cat...
lash out when
I stroke her?

Q Why are some cats so contrary? Take my cat. First, she jumps on to my lap and butts her head against my hand, asking to be petted. I stroke her, and she purrs away, seemingly completely relaxed and happy. Then, all of a sudden, she will turn and bite my hand quite viciously. She'll then leap off my lap as fast as she can, often scratching me in the process. I love to stroke her, and don't want to stop, but I am fed up of getting hurt. So, what should I do?

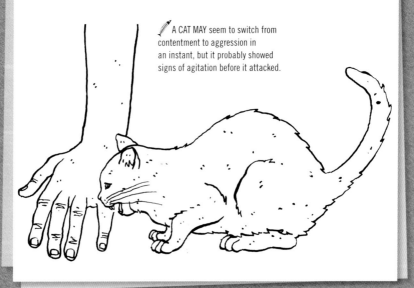

A CAT MAY seem to switch from contentment to aggression in an instant, but it probably showed signs of agitation before it attacked.

A This is certainly quite annoying for you, but it doesn't happen because your cat is "contrary," as you put it. **This is actually quite a common behavior pattern that has been dubbed "petting and biting syndrome."**

Cat behaviorists have several theories as to why some cats turn on the person who is petting them. First, it is possible that your cat, lulled by the pleasant sensation of being stroked, starts to drop off to sleep. She then wakes with a start, not quite knowing where she is, but aware that she is being confined (by your hands). So she instinctively struggles to get herself free.

Another possible explanation is that your cat enjoys only short periods of petting. You may not be picking up on her signs of discomfort. Be on the lookout for any twitching of the tail, growing restlessness, or a flattening of the ears. These are all signs that you should stop the petting. Also, avoid the sensitive stomach area: Many cats will lash out when this is touched.

Some cats will happily sit for long stretches in their owner's lap; others cannot bear to be held at all. Generally speaking, cats that are handled regularly between the ages of two and seven weeks tend to enjoy human contact, while kittens that do not have much exposure to human touch at this time shy away from it later in life. You can increase your cat's tolerance by handling her frequently for short periods, but you should always be sure to stop before she becomes agitated. Never strike a cat for biting; this will serve only to increase her anxiety at being handled.

CATS VARY CONSIDERABLY in how much they like to be petted. You will have to judge your own cat's tolerance for petting and build up sessions slowly.

SPECIES WATCH

It can be difficult to read the aggression signs of some animals. Take the great, wide yawn of the hippopotamus: It does not, as one might suppose, signal lazy contentment: It is actually a sign that it is about to attack. The animal is displaying the huge tusks that are his main weapon, and also giving his enemy a blast of his fearsomely stinky breath.

Why does my cat...
want to be let out as soon as I have let her in?

Q I have had three cats, and they have all been equally indecisive about coming in and going out of the house. I let them out, and five minutes later they want to come in again. So I let them in, and lo and behold, they want to go out again. What is it with all this to-ing and fro-ing? Why can't they just make their minds up about where they want to be? I can't get a cat flap because my home is rented.

AN INDOOR–OUTDOOR CAT wants to check its territory often, but may spend only short periods outside before returning to base.

A This is one of those cat–human conflict situations that can get in the way of a perfectly harmonious relationship. It seems to you that your cats are being quite unreasonable—surely they can stay out for a decent amount of time, then come in and stay in.

Your cats' apparent indecision can be explained by the fact that they are territorial animals, but that in a domestic situation, their territory is divided: Some of it is outside, and some of it inside. From your point of view, the front door is the beginning of your space; from your cats' point of view it is an unnatural barrier that runs through the middle of their domain.

Your cats want to patrol their territory regularly to check for the presence of other cats, and to refresh the scent signals that are their border-markers. This involves a certain amount of coming and going, and commits you to some tiresome opening and closing of doors. But if you think about it, you will probably realize that you are being asked to do this less often than you imagine. The occasions when you are repeatedly called upon to let your cats in or out naturally stick in your mind, but there are probably plenty of other times when your cats stay put for long periods.

So, in the end, it all boils down to a question of perception: Your cats' perception of their territory, and your perception of their demands on you as a kind of personal gatekeeper.

YOUR CAT IS not being indecisive when it goes in and out at frequent intervals. It is simply being a cat.

Chapter four
Solving problems

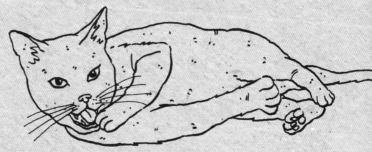

Cat owners tend to be loyal to their pets, and so may tolerate unappealing habits, such as soiling or spraying, for months, or even years. **But many problem behaviors are easily solved once you understand what lies behind them.** Some have medical causes, and need expert attention from a veterinarian in the first instance. Others are psychological: Cats are sensitive creatures that express their distress in a surprising variety of ways. Some behavioral problems can be stopped simply by identifying the source of stress and eliminating it. **Other matters may require more careful and complex handling.** The most severe conditions may even need the intervention of a specialist cat behaviorist. All of the problems discussed in this section offer glimpses of the unique character of the domestic cat, and so can help you to ensure that your animal leads the happiest, healthiest life possible.

Why does my cat...
keep spraying my walls?

Q My tomcat has started spraying in the hallway, creating the most horrendous smell. Every time I clean it up, he does it again. It is driving me crazy, especially as I have had the walls repapered and the carpet replaced recently. I have never caught him in the act, but I am sure that no other cat can get in our house (we don't have a cat flap, and my tom is kept indoors). I had him neutered young on the breeder's advice, to prevent this from happening. Why is he spraying—and, more importantly—how can I stop him from doing it?

SPRAYING INDOORS IS unusual behavior for a cat, and it is often a sign that the cat's sense of security has been disturbed.

A If there is one feline behavior that owners find hard to live with, it is spraying. Yet, for a cat, it is a completely natural way of marking out its territory. Unlike urination, spraying is done when the cat is standing up, and with its butt facing a vertical surface, such as a tree trunk. It takes this position so that the scent is deposited at nose height, where other cats are most likely to notice it. Although neutering stops most cats from spraying, some neutered cats still do it.

Spraying generally occurs out of doors. The home is a cat's safe base, so it usually does not need to spray here. It follows that indoor spraying indicates some kind of territorial insecurity. There are myriad causes—anything from the arrival of another cat to a dirty litter box can trigger spraying. It can also sometimes point to a medical problem, so is worth checking with a veterinarian. In your case, it is probably the fact that you have redecorated that has made your cat anxious: he is spraying your freshly papered walls to reaffirm that this is his territory.

A cat is more likely to spray somewhere that already smells of urine, so clean the area with an enzymatic cleaner or a dilute solution of biological laundry detergent. Then treat with rubbing alcohol, and let dry. (Do not use a cleaning product containing ammonia; this contains components also found in urine, and will encourage spraying.) Let your cat re-explore the area with you present. Once he has accepted it as his territory, the spraying should stop.

STAYING WITH YOUR cat while it explores the cleaned area will help to prevent it from spraying again.

Why does my cat...
eat woolly sweaters?

 I have a Burmese cat with one very strange habit: She eats woolen sweaters. She started sucking on my clothing when I was stroking her, then I noticed that she had chewed a hole in one of my sweaters. Now it is almost as if she is addicted. I put all woolen items away, but she has started to chew other fabric, too, including bed sheets and towels. Apart from the cost of replacing all the items she has ruined, I'm worried what eating this stuff may be doing to her insides. Why is she doing this, and what is the best way to handle it?

SUCKING MAY TURN into wool-eating. The habit may become so pronounced that the owner has to keep all clothing shut away.

A The first known case of a wool-eating cat was documented in the 1950s. Since then it has become clear that this behavior, albeit rather bizarre, is not uncommon. Curiously, it is more prevalent among some breeds than others. Siamese cats were the first to be reported as wool-eaters, but Burmese also seem particularly prone to it.

There are various theories as to why wool-eating happens; the most likely is that it is connected to the sucking behavior seen in cats who have been weaned too early (*see page* 24). This certainly seems to be the case with your cat, who, you say, started with sucking and then progressed to eating wool and other fabrics. The fact that fabric-eating is more common among Oriental breeds may be something to do with their sensitive temperament. It is also seen more often in cats that live indoors. Cats that are, by contrast, allowed to hunt prey outdoors seldom suffer from eating disorders.

Fabric-eating can be a serious problem. Cats that eat wool and other such materials are in danger of suffering from internal blockages. Possible treatments include playing with your cat more often, ideally several times a day, with, say, a feather on a string, so that she can "hunt." If your cat is kept indoors, consider whether you can give her access to the outside; this may help to reduce her dependence on you and increase the amount of stimulation that she receives. It may also help to give her naturally textured food that she has to tear apart with her teeth. See your veterinarian for advice.

WOOL IS NOT the only fabric that a cat will chew on. Its appetite may stretch to include towels, bedlinen, and other items.

Why does my cat...
keep running away from home?

Q Since we moved home a couple of months ago, our cat has repeatedly run away. Several times we have found him back at our old home, which is about 20 minutes' walk down the road. On the advice of a friend, we kept him indoors for a few days after the move, but he was very unhappy, and I am worried that it has just made him dislike our new home all the more. Our old neighbor has now offered to have him. Should we let her have our beloved cat, or can he adjust to our new home?

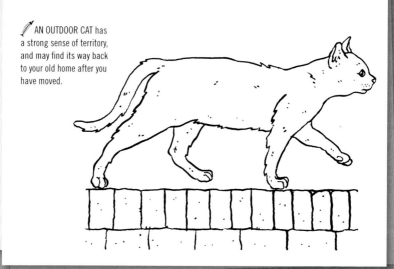

AN OUTDOOR CAT has a strong sense of territory, and may find its way back to your old home after you have moved.

A Moving house is stressful enough for humans. For a highly territorial animal such as a cat, the experience is much, much worse. It is not surprising that he hankers after all the old haunts. As you have moved just a mile or so, your cat is likely to come across familiar landmarks when he explores his new territory. Naturally enough, he takes the opportunity to navigate his way "home." Your old neighbor is clearly keen on cats, so it is likely that he meets a friendly reaction when he gets there—something that reinforces the message that his old home is a good place to be.

You don't need to give up on your pet yet. Your friend was right to say that cats should be confined after a house move, but a few days is not long enough. You should have kept him indoors for at least a month. You can still try this now. Settle him in one room, where he has food, water, and access to a litter box. Then gradually allow him access to the rest of the home, leaving the door to the room open at all times. **After four to six weeks, he should be relaxed and comfortable in the house.** Make sure that your outside area is securely fenced, then let him outside for a brief look around. It's good if he misses a meal beforehand; that way, he will be ready to come in when you signal that it is feeding time after a few minutes' exploring. Do this several times, before letting him out during the day only.

KEEPING A CAT indoors for several weeks after a house move will help it to adjust to the new environment more easily.

Why does my cat...
groom himself so hard that he gets a bald patch?

Q My sweet Abyssinian has a strange problem that I have never encountered before. He has started to groom himself so frequently that he has literally worn away some of the fur on his flank. I have tried stopping him from doing it, but I obviously can't watch him all the time, even though he is kept indoors. I am a new mom, so have probably been giving him less attention than he is used to. Is this the reason for his strange behavior?

EXCESSIVE GROOMING MAY be triggered by a traumatic event, and may then become a habitual behavior pattern. It needs careful handling.

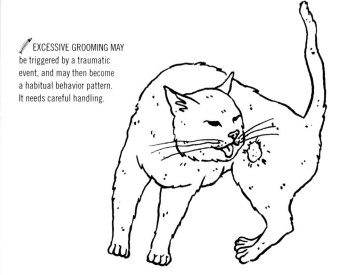

A It is quite normal for cats to spend hours grooming themselves, but when grooming is done so repeatedly that it causes baldness, there is clearly something wrong. Sometimes overgrooming can indicate the presence of fleas or other parasites, or it may be a sign of a medical problem. So the first thing that you should do is take your cat to the veterinarian for a checkup.

If no physical cause can be found, then his overgrooming is probably an obsessive behavior. Certain breeds of cat—including Siamese, Burmese, and Abyssinians—are particularly prone to it. It is important to deal with the habit early on, before it becomes fixed.

Excessive grooming is often a reaction to stress or trauma. There are many possible triggers: A house move, building works at home, or just boredom. As you are a new mom, it is likely that the arrival of the baby has disconcerted your cat. Consider whether changes were made to your cat's environment or routines because of the baby. Are you feeding him at different times, has his litter box or feeding station been moved, or has he lost his usual sleeping place? Minimize these changes as much as you can.

At the same time, prevent him from indulging in overgrooming as much as possible. Clap your hands to distract him, then give him lots of attention or play. Make sure that he has plenty of stimulation in the form of toys and other devices. If all else fails, your veterinarian may suggest a calming medication to help mitigate this distressing behavior.

REPEATEDLY BITING ONE paw, like overgrooming, is a type of obsessive behavior that is sometimes seen in cats.

SPECIES WATCH

Wild animals held in captivity are prone to obsessive behaviors—usually when the conditions in which they are held are less than ideal. Giraffes and camels will lick the walls incessantly; big cats will pace up and down in the same pattern until they wear a path in the ground; bears will chew the bars of a cage. In all cases, the treatment is to provide a better, more fulfilling environment.

Why does my cat...
no longer use his litter box?

Q My cat, who is six years old, has had exemplary toileting habits until now. A few weeks ago, she had an accident. I cleaned up the mess and thought no more about it, but then she did it again. Now she seems to go wherever she feels like it, and is not using the litter box at all. I am at my wit's end, and my home is starting to smell. How can I get her to be clean again?

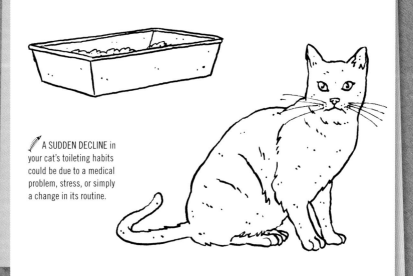

A SUDDEN DECLINE in your cat's toileting habits could be due to a medical problem, stress, or simply a change in its routine.

A The first thing to consider is whether your cat has a medical problem. Kidney disease or urinary infection can make it difficult for a cat to reach the box in time. So take your cat to the veterinarian for a checkup. If she is given a clean bill of health, consider whether there has been a change in her routine. Cats are highly sensitive to changes in their toileting environment. If you have had to move the box, for example, it could now be too close to lots of human activity. Using a new type of litter—or skimping on the litter—can also stop a cat using its box. Cats are extremely fastidious, and will not use a dirty box. Conversely, a cat will also refuse to use a box that smells too strongly of disinfectant or perfumed cleaner. Take steps to change anything that you identify as a problem.

If there is no physical reason for your cat's behavior, it may be that an unpleasant experience triggered her aversion to the litter box. If, say, your cat was scolded or given medication on the box, she may no longer want to go near it. In this instance, replacing the box or moving it can actually help. You can retrain your cat by confining her to the same room, and by gently placing her on the litter box after any accidents. Once she starts using the box again, you can gradually allow her access to the rest of the house.

Always clean up any accidents using a solution of biological laundry detergent, then spray with rubbing alcohol using a plant-mister. This removes any residual smell that might encourage the cat to view the area as their latrine in the future.

KEEPING A CAT in one room can help with retraining. No cat likes to soil near its eating area, so it is therefore more likely to use a carefully placed litter box.

Why does my cat...
attack my new boyfriend?

Q I have had my cat since he was a kitten. At the time, I was with my husband, but we have since divorced. My cat was a real solace to me when I was on my own, and I adore him. But he seems to hate my new partner. When the poor man is sleeping, my cat will pounce on his toes and bite them—hard. I put the cat outside the door, but he meowed continually, keeping us awake. Now the cat has started attacking my partner in the daytime, too. What can I do? Surely I don't have to choose between my cat and my man?

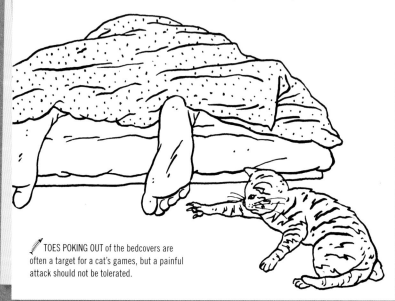

TOES POKING OUT of the bedcovers are often a target for a cat's games, but a painful attack should not be tolerated.

A Your cat certainly seems to be behaving like a jealous lover, and you are feeling powerless to mediate between these two "men" in your life. Don't worry: You can get the cat to accept the new situation, but it will take time, patience, and persistence.

Your cat is feeling threatened by your partner's arrival in your home. You need to show the cat that his attempts to control the situation are not working.

So, difficult as it may be, your partner should ignore the cat's attacks as much as he can. (Wearing thick clothing and stout shoes can help.) If your cat is unable to provoke a reaction, he will eventually stop. In the meantime, you, as well as your partner, must ignore his behavior totally: do not look at him, speak to him, or make any physical contact.

For the sake of your partner's toes, you have to put your cat out of the bedroom at night. Provide him with a comfortable alternative sleeping spot elsewhere, and leave some dried food out for him to nibble. Be prepared for some furious meowing for several nights, but do not give in. If you do, you will have to start the whole process again.

At the same time, give your cat lots of attention and playtime during the day. If you are holding your partner's hand, try petting your cat at the same time, so that he gets used to the idea that this isn't a "him or me" situation. Slowly, your cat will come to accept the situation. You may even find that he starts to go to your partner of his own accord.

A SHEEPSKIN HAMMOCK hung from a radiator makes a cozy bed for a cat, and may console it for being shut out of your bedroom.

Why does my cat...
scratch my furniture?

Q I got my beautiful cat as a kitten two years ago. She is an indoor cat who has adjusted just fine to living in my apartment, and has a wonderfully playful temperament. But she is a real scratcher. So far, she has ruined my couch and wallpaper, and she has even clawed my drapes. My colleague says that declawing is the only way to stop her from doing this. Is there another solution?

SCRATCHING IS ONE of the ways in which a cat will mark its territory. An indoor cat may scratch walls and drapes, as well as furniture.

A Your cat, like all felines, has a strong instinct to scratch. Outside, the cat would scratch trees, gateposts, sheds, and other suitable surfaces. **This entirely natural behavior serves not only to sharpen the cat's claws, but also to mark its territory.** The scratches leave a visual mark, and they contain the cat's scent, which is delivered from glands between the cat's paw pads.

So, your cat is just behaving as a cat should—she doesn't know that she is ruining your home. A solution must be found that both of you can live with. Declawing is definitely not the answer. People imagine that this procedure is the equivalent of trimming the cat's nails, but it actually involves removing the nail and first joint of the cat's "toes." There can be postoperative pain, and declawing can affect both the cat's balance and its ability to defend itself if it goes outside. For these reasons, declawing is illegal in some countries.

Instead, you must rechannel your cat's urge to scratch. Invest in a hessian scratching post. Choose one that is sturdy and does not wobble. It should be tall enough to allow your cat to stretch to her full height. Keep it near a favorite scratching place until she gets used to it; you can then move it to a less obtrusive area of the room. Be sure to let her explore it as she pleases— trying to show her how to scratch will just put her off. You should also ensure that she has plenty of toys, and that you play with her often to keep her stimulated.

SPECIES WATCH

There is only one member of the cat family that does not have retractable claws: the cheetah. The claws of the cheetah are always exposed. It is not clear why this should be, but it is a characteristic that contributes to the animal's astonishing speed. A cheetah's claws act like the cleats on a sprinter's shoes, gripping the ground as it runs, and driving it forward with every stride.

A SCRATCHING POST satisfies an indoor cat's need to scratch— and keeps its claws away from your furniture.

Why does my cat...
not want to use
his cat flap?

Q I have adopted a three-year-old indoor–outdoor cat from a friend. I work during the day, but have installed a cat flap so that he can get out into the backyard whenever he wishes. However, he steadfastly refuses to use it, preferring to wait until I come home and can let him out myself. I was advised to push him through in order to teach him that it is safe. But when I tried this, he just fought me and now seems to view it with real dislike. He loves being outdoors, and asks me to let him out as soon as I get home. So how can I get him to use the cat flap?

IT IS COMMON for a cat to shun the flap that its thoughtful owner has provided, and to demand to be let out in the usual way.

A The problem here is that your cat has a different idea of the cat flap than you do. You think that once your cat knows how easy it is to use a flap, he will love the freedom that it gives him. Your cat is confused by this contraption that propels him into the outside world without letting him check for enemies first. Worse still, his trusted human companion has tried to force him through it. So he has decided to steer clear of it.

You need to show him that the cat flap offers a safe route to the outdoors.

Start by propping the flap open using string (or strong adhesive tape), so that he can see the backyard beyond. When he is outside and asking to come in, encourage him to come through the flap by placing some food on the other side. Initially, it is easier to tempt a reluctant cat to come in rather than go out. Do this a few times, then try luring him out, using food or a favorite toy. Once he is familiar with the flap, you can lower it, inch by inch, over a period of a week or so. Your cat will eventually learn how to push the flap open. But do make sure that you don't rush him at all.

Be alert to other cats in the neighborhood accessing your home through the flap. If this happens, your cat's sense of security will be threatened. In these circumstances, it is best to board up the cat flap and let your cat in and out yourself—inconvenient though that may be for you.

TAPING THE FLAP open allows your cat to see that it offers a route to the outside. But it may take time before the cat will go through it.

Why does my cat...
refuse to go outdoors?

Q I have always kept my cat indoors because I lived in the city and was worried about her having an accident or getting into fights. Now, though, I have moved to a quiet neighborhood. I have a large yard, and I would like to give her the opportunity to go outdoors. I thought that she would jump at the chance, but she seems most reluctant. What is best: To keep her in, or to persevere with encouraging her to go out?

AN INDOOR CAT may choose not to go outside when given the opportunity. Do not force a cat to go outdoors if it is unwilling.

A There is great debate about this question. Many people in the USA keep their cats indoors. Research shows that indoor cats live much longer than those that are free to wander outside, where they face the dangers of disease, traffic, wild animals, and so on. Many rescue homes now insist that would-be owners keep their cat indoors.

Behavioral problems, such as scratching, spraying, and messing, are more likely to occur with indoor cats, suggesting that confinement can cause stress. **However, many veterinarians believe that cats adapt happily to an indoor life, provided that they get lots of attention and are encouraged to exercise.** Two cats living together seems to help indoor cats cope, though sharing a living space with many other cats can increase an individual's stress levels.

Since your cat is not showing any interest in the outdoors, it may be best to let her continue as an indoor cat. Or, you could try giving her a taste of the outdoors through "supervised outings." Make sure that your yard is well fenced so that your cat cannot escape. Then carry her out into the garden. She may prefer to stay in your arms, but you could also try letting her explore while you keep watch. If she goes near the fence, clap your hands to dissuade her. Keep outings short, so that she is not tempted to go farther afield—20 minutes or so should be sufficient.

SUPERVISED OUTINGS ALLOW a nervous cat to experience the sights, smells, and sounds of the outside world in safety.

Why does my cat...
prefer pond water to fresh?

Q Every day I put out fresh water for my cat and leave it in a bowl by her food. But she invariably turns up her nose at it, and chooses to drink from our pond instead. I have even seen her drink from a muddy puddle—gross. I'm worried that she will catch something, though she seems okay so far and has been doing this for some years. Why does she prefer stagnant water to fresh?

THE WATER IN a pond or puddle is more attractive to some cats than water from the faucet, which may smell of chlorine.

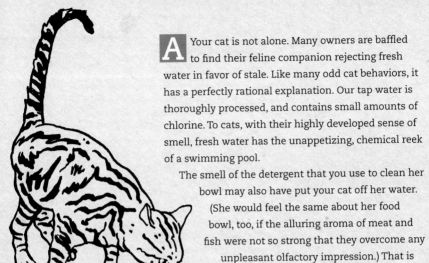

A Your cat is not alone. Many owners are baffled to find their feline companion rejecting fresh water in favor of stale. Like many odd cat behaviors, it has a perfectly rational explanation. Our tap water is thoroughly processed, and contains small amounts of chlorine. To cats, with their highly developed sense of smell, fresh water has the unappetizing, chemical reek of a swimming pool.

The smell of the detergent that you use to clean her bowl may also have put your cat off her water. (She would feel the same about her food bowl, too, if the alluring aroma of meat and fish were not so strong that they overcome any unpleasant olfactory impression.) That is why she goes and drinks from the pond. To your cat, the smell of stagnant or muddy water is, above all, natural.

Since your cat seems to suffer no ill effects, there is probably no reason to worry—in the wild, animals would certainly drink from puddles and ponds. However, because some diseases can be transmitted through dirty water, it is best to encourage your cat to drink fresh. Make sure that you rinse your cat's dishes very well, and offer her natural spring water to drink.

Cats tend to need less water than other animals, but they do seem to be particular about their drinking habits. Some cats prefer drinking out of a cup or glass rather than a bowl. Others insist on running water, and like to drink only from the toilet or a dripping faucet. Yet others like to drink, rather charmingly, by dipping a paw in the water and sucking it.

RINSING THE WATER bowl thoroughly will help to remove the smell of detergent that most cats find so repellent.

Why does my cat...
yowl at night, when he never used to?

Q My cat has recently started yowling during the night. Some time after midnight, he will meander through the house, making an awful howling noise. I go to find him, and he looks as if nothing has happened. Usually, I give him a cuddle, and take him back to my bed, where he snuggles down to sleep. He's 14 years old and doesn't seem to be ill or hurt, and he is fine during the day. So is he trying to tell me something?

YOWLING AT NIGHT is common in older cats, and may be a habit that has developed in order to gain the owner's attention.

A First of all, nighttime yowling can be an indicator of various health problems, including hyperthyroidism or high blood pressure. So, in the first instance, you should take your cat to your veterinarian to check that he is not ill.

Many healthy cats start to yowl at night as they get older. It could be that weakening eyesight and hearing make negotiating the house at night more difficult. Confused or disorientated, the cat yowls. Another possible cause is that the cat feels lonely. He could come to find you, but now that he is older, he finds it easier to ask you to come to him.

Younger cats can also yowl at night. The behavior is usually triggered by something that makes them feel insecure, such as a house move or the arrival of a new baby. The yowling can quickly become habitual if the cat finds that it gets your attention.

Most owners are kind enough to indulge nighttime yowling in an older cat. But you will probably want to retrain a younger cat. To do this, you need to be a little hard-hearted. Shut your bedroom door, equip yourself with earplugs, and ignore the yowling, however piteous it becomes. Eventually, the cat will learn that yowling doesn't work. It's best to combine this approach with some kind of aversion therapy, such as making a loud noise when the yowling begins. But don't embark on this approach unless you are sure that there is no medical cause for the yowling, and only if you know that your cat is safe.

IF THE CAUSE of the yowling is loneliness, the cat may settle down to sleep as soon as it is near you.

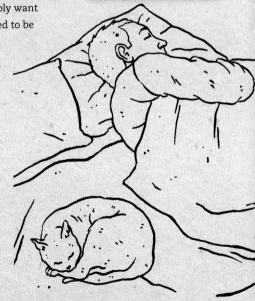

SPECIES WATCH

It is commonly believed that wolves howl at the moon (especially a full one). In fact, wolves howl at night, whether the moon is up or not. The howling is a bonding activity for the wolf pack, and also a way of advertising the pack's presence to other clans. From a human perspective, the protracted howl of the wolf happens to be one of the most eerie and chilling sounds in nature.

Why does my cat...
beg for food all the time?

 I have adopted a beautiful, two-year-old stray from the rescue center. I am really enjoying his company, except for one thing: Whenever I am in the kitchen, he spends the whole time crying for food. He jumps up on to the work surfaces when I am cooking and on to my lap when I am trying to eat. He has three meals a day, and I do give him the odd scrap from my dinner as well. Why is he acting so hungry all the time?

CONSTANT BEGGING IS an annoying habit. Your cat will only stop doing it if you refuse to reward the begging with food.

A It's great that you have given a rescue cat a loving home. Now you and he have to find a way of living together. While he was a stray, he may have gotten into the habit of scrounging for food all the time, because he didn't know where his next meal was coming from. This behavior has become ingrained, and you have inadvertently reinforced the message that it works by giving him scraps from your plate.

You clearly need to set some boundaries for your cat, but first consider whether he is a normal weight for his size. Strays sometimes need extra food for a while until they reach a stable weight. Take him to the veterinarian to ensure that he is not underweight or unwell.

If you are sure that your cat is getting enough food, then you need to teach him that begging simply doesn't work. The only way to do this is to refuse to reward this behavior. Feed him at set times rather than on demand. And never feed him from your plate: If you do this even occasionally, he will think that he has a right to your food.

Do not allow him on to your work surfaces. Every time he jumps up, simply take him off. If he persists, shut him out of the kitchen for a short period to help him get the message. Remain calm at all times; shouting at a cat will make it stressed, and is likely to cause more behavioral problems than it solves. In time, your cat will realize that begging does not reap rewards, and he will stop.

SPECIES WATCH

These days, stray cats will often find themselves in competition with other urban scavengers, such as foxes and sea gulls. Foxes will eat anything, and have learned to open trash cans and Dumpsters to get at the edible garbage within. Sea gulls find easy pickings on landfill sites, but are sometimes bold enough to swoop on a person eating an ice cream and snatch it from their hand.

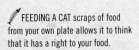
FEEDING A CAT scraps of food from your own plate allows it to think that it has a right to your food.

Why does my cat...
fight when it is time
to go to the veterinarian?

Q My three-year-old queen cannot bear going to the veterinarian. If she even sees the cat carrier that I use to take her there, she dashes off and hides. On several occasions, I have had to cancel the appointment because I can't find her. She has also scratched me quite badly when I have caught her unawares and tried to put her in the carrier. It's got to the point where I have started to dread the appointments as much as she does. How can I make this an easier experience for both of us?

GETTING AN UNWILLING cat into a carrier can be fraught with tension, and you may end up with a nasty scratch.

A You are not alone in this problem—countless cat owners have found that their cats absent themselves when they are due at the veterinarian's. It's hardly surprising that they dislike going to the veterinarian, since this involves going to a strange, new environment where there may be hostile cats, as well as other animals. And they are often subjected to incomprehensible and unpleasant experiences such as examinations or injections. If the carrier is used only for veterinarian visits, there is a clear association with a negative experience in the cat's mind. As soon as it sees the carrier, it makes a run for it.

To get your cat to the veterinarian, be sure that you are calm yourself—if your cat picks up on the fact that you are stressed, this is likely to make her anxious. Prepare the carrier out of sight (and hearing), then open the door and place it with the back against the wall so that it is stable when you move the cat into it. Pick up your cat, holding her securely, and take her to the carrier. Put her in butt-first—not head-first; place her butt and back legs in the carrier as you push gently on the cat's head to move her into the carrier. Then close the door—quickly.

If your cat is very nervous, it is helpful to leave the carrier out all of the time. Place a blanket inside, so that it is comfortable, and let your cat find some dry food or a catnip toy there. This will help her build positive associations with the carrier, and should make it easier to get her into it when necessary. One or two drops of lavender or chamomile oil on a cotton ball placed in the carrier can also calm your cat.

IF THE CAT associates the carrier with going to the veterinarian, it will dash off as soon as the hated item makes an appearance.

Why does my cat...
seem unhappy now that he has a companion?

Q My friend moved overseas and asked me to give a home to her much-loved cat, who is about the same age as mine (both are neutered males used to living indoors). They seem to get along just fine. However, I have realized that I now see more of my friend's cat than I do of my own. He has taken to spending much of the day in my bedroom. He comes downstairs to eat and use the litter box, but doesn't seem to hang out with me any more. Another thing is that although he still defecates in the litter box, he has made quite a few puddles elsewhere. What's the reason for this?

BLOCKING THE ROUTE to the food, litter box, or sleeping area is an effective way for one cat to bully another.

A Owners like to think that their cats get along. But just because they don't fight, that does not mean that there is no struggle going on. The fact that your cat cowers in the bedroom all day means that there is an uneasy relationship. And as it is your cat who has retreated from the main part of the house, your friend's cat is obviously the more dominant of the two.

A dominant cat may block another cat's access to its latrine or feeding place as a way of bullying the weaker one into submission. As your cat is suddenly having accidents, it could be that your friend's tom is intimidating him in this way. Your cat manages to defecate in the box because it is easier to hold on to a bowel movement. But he needs to urinate more frequently, so is forced to go where he can.

To help solve the problem, get at least two more litter boxes and position them in different places around the home (including at least one near your bedroom). Since your friend's cat won't be able to patrol them all, he will not be able to control your cat's access. For the same reason, make sure that the cat can escape from the litter box easily; don't wedge it into a corner. You should also provide separate feeding stations and beds. To increase the amount of territory for your cats to explore, install high perches, such as cat trees, and invest in scratching posts, as well as toys. This should help your cats to coexist more happily. If there is no improvement, you may need to consider rehoming your friend's cat.

CAT TREES SATISFY a cat's desire to be up high, and also increase the amount of space that it has to explore.

Index

Acknowledgments

With thanks to all the cat lovers who shared their experiences with me, especially Clare Lanchbery, Jacqui Boddington, Juliet Cox, Wenda and Allan Bradley, and Jonathan Bastable. And a special mention to Cleo, Marmaduke, Daisy, Frank, Nancy, Sid, and Oscar.